新形态教·学·练
一体化规划丛书

Android Studio App 边做边学

微课视频版

◎ 刘韬 郑海昊 编著

清华大学出版社
北京

内 容 简 介

本书以开发和设计完整的 App 为导向，采用微课项目教学的方式组织内容，项目之间彼此承接与演进。本书内容主要涵盖了 10 个项目，分别从 App 概念及应用、如何搭建 Android 开发环境、如何在 Android Studio 平台下设计与开发 App 的 UI 界面（包括添加 UI 组件、多媒体开发、数据传递与多线程）以及如何打包发布等方面进行讲解。在每个项目的具体实施中，包括项目目标、项目准备、项目运行、项目结案及项目练习五部分。

本书适合作为高等院校相关专业，尤其是数字媒体、网络与新媒体、游戏和动漫等专业的教、学、做一体化教材。

本书封面贴有清华大学出版社防伪标签，无标签者不得销售。

版权所有，侵权必究。举报：010-62782989，beiqinquan@tup.tsinghua.edu.cn。

图书在版编目(CIP)数据

Android Studio App 边做边学：微课视频版/刘韬，郑海昊编著. —北京：清华大学出版社，2020
（2022.7重印）
（21世纪新形态教·学·练一体化规划丛书）
ISBN 978-7-302-53617-8

Ⅰ.①A… Ⅱ.①刘… ②郑… Ⅲ.①移动终端－应用程序－程序设计 Ⅳ.①TN929.53

中国版本图书馆 CIP 数据核字(2019)第 173909 号

策划编辑：魏江江
责任编辑：王冰飞
封面设计：刘　键
责任校对：时翠兰
责任印制：曹婉颖

出版发行：清华大学出版社
网　　址：http://www.tup.com.cn，http://www.wqbook.com
地　　址：北京清华大学学研大厦 A 座　　　邮　编：100084
社　总　机：010-83470000　　　　　　　　　邮　购：010-62786544
投稿与读者服务：010-62776969，c-service@tup.tsinghua.edu.cn
质量反馈：010-62772015，zhiliang@tup.tsinghua.edu.cn
课件下载：http://www.tup.com.cn，010-83470236

印 装 者：三河市铭诚印务有限公司
经　　销：全国新华书店
开　　本：203mm×260mm　　印　张：15　　字　数：283 千字
版　　次：2020 年 5 月第 1 版　　　　　　　印　次：2022 年 7 月第 4 次印刷
印　　数：4001～5000
定　　价：49.80 元

产品编号：084334-01

FOREWORD 前 言

App 是移动多媒体时代重要的交流工具,是设计学与计算机科学等相关专业人才走向社会急需掌握的一项技术,相关的开发与设计是 Android 开发技术人员需要掌握的基本技能。在 App 开发的过程中,对 UI 的设计是十分重要的,也是 App 能否被用户接受和持续使用的关键性指标。因此,在高等院校艺术设计类专业中开设 App 开发与设计的相关课程既满足市场需求,又符合行业发展趋势。App 已成为一门重要的专业课程。本书以实例的方式,通过项目引导训练读者在 Android Studio 平台下开发与设计 UI 界面的技术手段与实现方法。

本书以开发和设计完整的 App 为导向,采用微课项目教学的方式组织内容,项目之间彼此承接与演进。本书内容主要涵盖了 10 个项目,分别从 App 概念及应用、如何搭建 Android 开发环境、在 Android Studio 平台下如何设计与开发 App 的 UI 界面(包括添加 UI 组件、多媒体开发、数据传递与多线程)以及如何打包发布等方面进行讲解。在每个项目的具体实施中,包括项目目标、项目准备、项目运行、项目结案及项目练习五部分。项目目标部分给出设计与开发的任务,即读者需要掌握哪些知识和能够实现什么样的效果;项目准备部分用理论说明的方式介绍项目实现的技术方法和操作思路;项目运行部分介绍设计与开发 App 的实际案例,通过细致的过程演示,图文并茂地实现项目目标;项目结案部分对项目进行汇总式说明,总结项目中用到的技术知识点等;项目练习部分则是围绕项目需要掌握的重点,精心提供了适量的习题,供读者检测学习效果。

通过 10 个项目的具体学习和训练,读者不仅能够掌握使用 Android Studio 开发 App 的技术要点,而且能够体验到 App 开发中 UI 设计的重要性,项目内容的设置尤其适合数字媒体艺术、数字媒体技术、网络与新媒体、游戏设计等专业人才在具备艺术设计与多媒体创作的基础上进行技术技能方面的补充。

本书的参考学时为 48~68 学时,建议采用理论实践一体化教学模式,各项目的参考学时见下面的学时分配表。

学时分配表

项 目	课 程 内 容	学 时
项目 1	了解 App 的前世今生	2～4
项目 2	搭建 Android 开发环境	4～6
项目 3	创建第一个 App	2～4
项目 4	设计 App 的用户界面	6～8
项目 5	理解 App 的活动	4～6
项目 6	设置 App 的 UI 组件	8～10
项目 7	设置 App 的多媒体应用	8～10
项目 8	设置 App 的图像与动画	8～10
项目 9	获取 App 的数据	4～6
项目 10	发布 App	2～4
课时总计		48～68

注：本书提供 150 分钟的视频讲解，扫描书中相应位置的二维码，可以在线观看学习；本书还提供部分实例的程序源码，扫描目录上方的二维码可以下载；本书还提供教学大纲、教学课件、电子教案、习题答案、教学进度表等配套资源，扫描封底的课件二维码可以下载。

本书由刘韬、郑海昊编著，刘韬编写了项目 1、项目 2、项目 3 和项目 10，其他项目均为郑海昊编写。杨豪同学负责提供本书所用图片及视频讲解，乔露同学负责制作本书的 PPT。另外，特别感谢张荣老师对本书的指导与建议。

由于编者水平和经验有限，书中难免有欠妥之处，恳请读者批评指正。

编 者

2020 年 1 月

CONTENTS 目 录

源码等资源下载

项目1　了解 App 的前世今生 ……………………………………………………………… 1

 1.1　项目目标：了解 App 相关的基本概念与应用 ………………………………… 1
 1.2　项目准备 ………………………………………………………………………… 2
  1.2.1　手机系统的发展 ………………………………………………………… 2
  1.2.2　手机关键技术的发展 …………………………………………………… 5
  1.2.3　App 的技术特点 ………………………………………………………… 8
  1.2.4　Android 开发工具介绍 ………………………………………………… 8
 1.3　项目运行 ………………………………………………………………………… 10
  1.3.1　App 分类及应用 ………………………………………………………… 10
  1.3.2　App 的发展趋势 ………………………………………………………… 11
 1.4　项目结案 ………………………………………………………………………… 13
 1.5　项目练习 ………………………………………………………………………… 13

项目2　搭建 Android 开发环境 …………………………………………………………… 14

 2.1　项目目标：搭建设计 App 的 Android 开发环境 …………………………… 14
 2.2　项目准备 ………………………………………………………………………… 15
  2.2.1　Android 的体系介绍 …………………………………………………… 15
  2.2.2　Android 的安装文件介绍 ……………………………………………… 17
 2.3　项目运行 ………………………………………………………………………… 19
  2.3.1　安装 JDK ………………………………………………………………… 19
  2.3.2　安装 Android Studio …………………………………………………… 24

     2.3.3　配置 Android SDK ······ 28

     2.3.4　安装 AVD ······ 30

     2.3.5　Android Studio 操作指南 ······ 33

  2.4　项目结案 ······ 55

  2.5　项目练习 ······ 56

项目3　创建第一个 App ······ 57

  3.1　项目目标：用 Android Studio 创建 App ······ 57

  3.2　项目准备 ······ 57

     3.2.1　Android 的内部结构 ······ 57

     3.2.2　Android 的开发流程 ······ 59

  3.3　项目运行 ······ 59

     3.3.1　创建一个 App ······ 59

     3.3.2　运行 App ······ 60

     3.3.3　调试 App ······ 67

  3.4　项目结案 ······ 69

  3.5　项目练习 ······ 69

项目4　设计 App 的用户界面 ······ 70

  4.1　项目目标：通过视图创建 App 的用户界面 ······ 70

  4.2　项目准备 ······ 70

     4.2.1　介绍视图类 ······ 71

     4.2.2　介绍资源文件夹 ······ 72

     4.2.3　介绍布局类 ······ 79

  4.3　项目运行 ······ 82

     4.3.1　字符串资源 ······ 82

     4.3.2　颜色资源 ······ 84

     4.3.3　尺寸资源 ······ 85

     4.3.4　图片资源 ······ 86

     4.3.5　布局类 ······ 89

4.4　项目结案 ··· 93

　　4.5　项目练习 ··· 93

项目5　理解 App 的活动 ··· 94

　　5.1　项目目标：理解 App 的活动机制与状态 ································· 94

　　5.2　项目准备 ··· 94

　　　　5.2.1　介绍 Activity 的状态 ·· 95

　　　　5.2.2　介绍 Activity 的生命周期 ······································· 96

　　　　5.2.3　介绍 Activity 的属性 ·· 98

　　5.3　项目运行 ·· 100

　　　　5.3.1　创建新的 Activity ··· 101

　　　　5.3.2　为新建 Activity 设置属性 ······································ 104

　　　　5.3.3　启动 Activity ·· 105

　　　　5.3.4　关闭 Activity ·· 105

　　5.4　项目结案 ·· 106

　　5.5　项目练习 ·· 107

项目6　设置 App 的 UI 组件 ·· 108

　　6.1　项目目标：添加与设置 App 的 UI 组件 ······························ 108

　　6.2　项目准备 ·· 108

　　　　6.2.1　介绍 UI 组件：TextView 及其子类 ························· 108

　　　　6.2.2　介绍 UI 组件：ImageView 及其子类 ······················· 113

　　　　6.2.3　介绍 UI 组件：AdapterView 及其子类 ···················· 114

　　　　6.2.4　介绍 UI 组件：ProgressBar 及其子类 ····················· 117

　　　　6.2.5　介绍 UI 组件：ViewAnimator 及其子类 ·················· 121

　　6.3　项目运行 ·· 122

　　　　6.3.1　在 UI 中设计文本框：TextView 组件实例 ················ 122

　　　　6.3.2　在 UI 中设计可编辑文本框：EditText 组件实例 ········ 123

　　　　6.3.3　在 UI 中设计计时器：Chronometer 组件实例 ············ 128

　　　　6.3.4　在 UI 中设计单选按钮：RadioGroup 组件实例 ·········· 131

　　　　6.3.5　在 UI 中设计显示图片：ImageView 组件实例 ··········· 134

6.3.6 在 UI 中设计列表：ListView 组件实例 ……………………… 135
6.3.7 在 UI 中设计列表选择框：Spinner 组件实例 ……………………… 137
6.3.8 在 UI 中设计网格视图：GridView 组件实例 ……………………… 138
6.3.9 在 UI 中设计进度条：ProgressBar 组件实例 ……………………… 142
6.3.10 在 UI 中设计滑动条：SeekBar 组件实例 ……………………… 144
6.3.11 在 UI 中设计星级评价条：RatingBar 组件实例 ……………………… 146
6.3.12 在 UI 中设计图片查看器：ImageSwitcher 组件实例 ……………………… 149
6.4 项目结案 ……………………… 152
6.5 项目练习 ……………………… 152

项目7 设置 App 的多媒体应用 ……………………… 153

7.1 项目目标：为 App 添加多媒体应用 ……………………… 153
7.2 项目准备 ……………………… 153
    7.2.1 介绍音频控制类 ……………………… 153
    7.2.2 介绍视频控制类 ……………………… 155
    7.2.3 介绍相机控制类 ……………………… 156
7.3 项目运行 ……………………… 157
    7.3.1 设计音频控制 ……………………… 157
    7.3.2 设计视频控制 ……………………… 165
    7.3.3 设计相机控制 ……………………… 171
7.4 项目结案 ……………………… 176
7.5 项目练习 ……………………… 176

项目8 设置 App 的图像与动画 ……………………… 177

8.1 项目目标：为 App 添加或设置图像与动画 ……………………… 177
8.2 项目准备 ……………………… 177
    8.2.1 介绍绘图类 ……………………… 177
    8.2.2 介绍图像特效 ……………………… 181
    8.2.3 介绍动画类型 ……………………… 182
8.3 项目运行 ……………………… 185

　　　　8.3.1　添加图形图像　…… 185

　　　　8.3.2　设计图像特效　…… 189

　　　　8.3.3　设计动画　…… 195

　8.4　项目结案　…… 202

　8.5　项目练习　…… 203

项目9　获取 App 的数据　…… 204

　9.1　项目目标：获取数据与线程设置　…… 204

　9.2　项目准备　…… 204

　　　　9.2.1　介绍多线程　…… 204

　　　　9.2.2　介绍消息类　…… 206

　　　　9.2.3　介绍消息处理类　…… 206

　9.3　项目运行　…… 207

　　　　9.3.1　创建一个线程　…… 207

　　　　9.3.2　添加消息类　…… 211

　　　　9.3.3　添加消息处理类　…… 215

　9.4　项目结案　…… 217

　9.5　项目练习　…… 218

项目10　发布 App　…… 219

　10.1　项目目标：打包与发布 App　…… 219

　10.2　项目准备　…… 219

　　　　10.2.1　介绍 META-INF 文件夹　…… 219

　　　　10.2.2　介绍 jar 包与 arr 包　…… 221

　　　　10.2.3　介绍 App 如何上线　…… 222

　10.3　项目运行　…… 223

　10.4　项目结案　…… 226

　10.5　项目练习　…… 226

附录 A　综合案例　…… 227

了解App的前世今生

1.1 项目目标：了解 App 相关的基本概念与应用

App(Application)，即应用程序，多指在智能手机上安装的移动应用软件。在智能媒介蓬勃发展的今天，手机作为"人的延伸"已经与人的生活、工作、学习、娱乐等方方面面形成了无缝连接的状态。尤其在中国，App 的应用范围几乎涵盖了整个社会，各行各业都将发展的视野投放到了这个入门快、上手易、传播广的应用程序上。因此，App 的开发与设计对于学习软件开发或艺术设计等专业的学生来说是必不可少的，也是未来就业的基础敲门砖。

App 发展至今，种类十分庞杂，应用领域五花八门，提供的服务多种多样。但究其根本功能都与"表达"与"交流"息息相关。因此，"交互"的功能实现与设计就成为 App 开发的核心与重点，也是本书重点要讲授的内容。

本项目的项目目标为介绍与 App 相关的重点概念与应用，包括 App 运行的系统基础，即手机系统的发展概况，手机关键技术的发展，App 的界面设计，App 开发工具等内容。通过本项目的讲解，希望能够让大家对 App 的前世今生形成一个完整的概念，为后续开发、设计的学习提供良好的全局观。

1.2 项目准备

1.2.1 手机系统的发展

手机应用的升级往往伴随着系统的不断演进,不论是硬件上提高用户体验的操控感,抑或是软件系统的研发革新,都从不同的角度为用户带来了更为智能化的移动生活。几款主流的操作系统也一直处于战国纷争、势均力敌的状态。与此同时,新兴系统也在庞大的移动终端市场中渐入佳境、不断壮大。手机软硬件系统的多样化是目前智能手机市场发展的主要特点之一。下面将介绍几款主流的手机硬件系统与软件操作系统,并梳理其发展的历程与态势。

1. 手机硬件系统

随着通信产业的不断发展,移动终端的硬件系统演变也是一日千里。从最早的"一只顺风耳"摇身一变,出落成集话音、数据、图像、音乐、视频、动画、网页等多媒体于一身的"万花筒"。总的来说,手机的硬件系统的组成部分一般包括屏幕、CPU、GPU、话筒、听筒、摄像头、重力感应、无线连接、蓝牙、外部存储器等。

手机的硬件系统中与用户关系最为密切的就是屏幕了。纵观近几年手机市场的发展,厂家不遗余力地推出各种超大屏幕、显示超清、触感灵敏的款式,将用户对于手机通话的"功能性依赖"转移到以视像为主的"娱乐性依恋"。6英寸甚至更大屏幕的智能手机频频问世,AMOLED、IPS等面板技术继续引领高端配置,让用户体验顶级的色彩和流畅的触控感,同时也为更好方案的出现进行铺垫。

GPU是智能手机中的另一个亮点,是对数字图像线上线下的处理能力。随着四核处理器在手机领域的应用普及,用户体验到了更强劲的运行性能,能在手机上实现多媒体信息的处理工作,同时,也掀起新一轮产品内部架构设计的革新。

2. 手机软件系统

手机硬件系统跃步前进的同时,软件操作系统也在如火如荼地发展。相对于硬件,操作系统的升级是悄然进行的,它并不是以强烈的视觉冲击带给用户全新体验,而是通过兼容并衍生更多的应用程序来丰富手机用户的操作体验。

iPhone 的成功使行业内部达成共识。一方面,各大手机厂商已经达成了默契,即界面的操作方式与菜单列表的形式趋于统一,打破不同操作系统之间的隔阂;另一方面,手机市场又呈现出多样化的特点。主流的操作系统如 Android、iOS 等都各自占领着部分市场。根据凯度移动通信消费者指数(Kantar Worldpanel ComTech)的智能手机操作系统数据调查显示,2019 年第二季度,Android 操作系统占比最高,为 77.14%,高居榜首,与上季度占比基本一致;iOS 操作系统占比 22.83%,位居第二;而其他移动端操作系统相加占比才 0.04%,完全不及安卓和 iOS 的零头。从品牌占安卓手机销售的百分比来看,华为手机在中国城市中的销量继续占据主导地位,达到 36%,坐稳了中国市场份额第一的位置。

1) iOS

iOS 是由苹果公司为 iPhone 开发的操作系统。它主要是给 iPhone、iPod touch 以及 iPad 使用的。原本这个系统被命名为 iPhone OS,直到 2010 年 6 月 7 日,在 WWDC 大会上被改为 iOS。iOS 的系统架构分为 4 个层次:核心操作系统层(the Core OS layer)、核心服务层(the Core Services layer)、媒体层(the Media layer)和可轻触层(the Cocoa Touch layer)。操作系统占用大概 240MB 的存储器空间。iOS 操作系统界面如图 1-1 所示。

图 1-1　iOS 操作系统

2) Android

Android(安卓,其操作系统的 LOGO 如图 1-2 所示)是 Google 开发的基于 Linux 平台的开源手机操作系统。它包括操作系统、用户界面和应用程序,即移动电话工作所需的全部软件,而且不存在以往任何阻碍移动产业创新的专有权障碍。Google 与开放手机联盟合作开发了 Android,这个联盟由包括中国移动、摩托罗拉、高通、宏达电子和 T-Mobile 在内的 30 多家技术和无线应用的领军企业组成。Google 通过与运营商、设备制造商、开发商和其他有关各方结成深层次的合作伙伴关系,希望借助建立标准化、开放式的移动电话软件平台,在移动产业内形成一个开放式的生态系统。

3) Windows Phone

Windows Phone(其操作系统的 LOGO 如图 1-3 所示)的推出是微软进军移动设备领域的重大品牌调整,它包括 Pocket PC、Smartphone 和 Media Centers 三大平台体系,均面向个人移动电子消费市场。Pocket PC 针对无线 PDA 设计;Smartphone 专为手机设计,已有多个来自 IT 业的新手机厂商使用,市场占有率增长较快;Media Centers 是手机的媒体中心,为手机提供媒体集成储备功能。

图 1-2　Android 操作系统的 LOGO

图 1-3　Windows Phone 操作系统的 LOGO

Windows Phone 为手机开发提供的功能非常多,在不同的平台上实现的功能互有重叠也各有侧重。这 3 个平台都支持与台式机数据同步。例如,Pocket PC 的功能侧重个人事务处理和简单的娱乐,主要支持的功能有日程安排、移动版 Office、简单多媒体播放。Smartphone 提供的功能侧重于联系方面,它主要支持的功能有电话、电子邮件、联系人、即时消息。

4) Symbian

Symbian(塞班)是一个实时性、多任务的纯 32 位操作系统,具有功耗低、内存占用少等特点,非常适合手机等移动设备使用,经过不断完善,可以支持 GPRS、蓝牙、SyncML 以及 3G 技术。最重要的是,作为一个标准化的开放式平台,任何人都可以为支持 Symbian 的设备开发软件。与微软公司产品不同的是,Symbian 将移动设备的通用技术,也就是操作系统的内核,与图形用户界面技术分开,能很好地适应不同输入方式的平台,也可以为厂商提供制作友好操作界面的机会,迎合个性化的潮流。因此,用户才能见到显示方式迥然不同的 Symbian 系统。

5) 其他操作系统

在移动通信领域中,还有其他一些操作系统,如高通公司推出的 BREW 平台(它是基于 CDMA 网络"无线互联网发射平台"上的增值业务而开发的基本平台),源代码开放的操作系统 Linux,融云数据存储、云计算服务和云操作系统为一体的新一代操作系统阿里云 OS 等,种类繁多,但用户量较少。

1.2.2 手机关键技术的发展

智能手机的发展离不开信息技术的发展与革新。3G 移动通信技术和 4G 移动通信技术是对手机发展影响深远的代表性、关键性技术。正是这两大关键技术才使得智能手机真正成为人们生活、工作、学习、娱乐的亲密伴侣,也使得视频、游戏等大数据量的媒体资源能够实时上传或下载,扩展了手机的可视化功能。

1. 3G 移动通信

第三代移动通信技术(3rd Generation,3G)是指将无线通信与国际互联网等多媒体通信结合的第三代移动通信系统。相对于第一代模拟制式手机(1G 手机)和第二代 GSM、TDMA 等数字手机(2G 手机),使用 3G 技术的手机能够处理图像、音乐、视频流等多种媒体形式,提供包括网页浏览、电话会议、电子商务等多种信息服务,如图 1-4 所示。为了提供这种服务,无线网络必须能够支持不同的数据传输速率,也就是在室内、室外和行车的环境中能够分别支持至少 2Mb/s、384kb/s 以及 144kb/s 的传输速率。

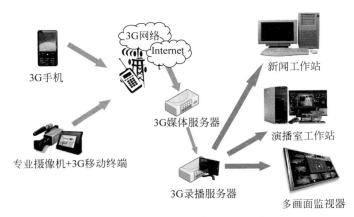

图 1-4　3G 网络的应用范围及服务类型

国际电信联盟(ITU)在 2000 年 5 月确定了 WCDMA、CDMA 2000、TD-SCDMA 三大主流无线接口标准,并写入 3G 技术指导性文件《2000 年国际移动通信计划》(简称 IMT-2000);2007 年,WiMAX 也被接受为 3G 标准之一。码分多址(Code Division Multiple Access,CDMA)是第三代移动通信系统的技术基础。第一代移动通信系统采用频分多址(Frequency Division Multiple Access/Address,FDMA)的模拟调制方式,这种系统的主要缺点是频谱利用率低,信令干扰话音业务。第二代移动通信系统主要采用时分多址(Time Division Multiple Access,TDMA)的数字调制方式,提高了系统容量,并采用独立信道传送信令,使系统性能大大改善。但 TDMA 的系统容量仍然有限,越区切换性能仍不完善。CDMA 系统以其频率规划简单、系统容量大、频率复用系数高、抗多径能力强、通信质量好、软容量、软切换等特点显示出巨大的发展潜力。下面分别介绍 3G 的几种标准。

1) WCDMA

WCDMA 全称为 Wideband CDMA,也称为 CDMA Direct Spread,意为宽频分码多重存取,是基于 GSM 网发展出来的 3G 技术规范,是欧洲提出的宽带 CDMA 技术,它与日本提出的宽带 CDMA 技术基本相同。WCDMA 的支持者主要是以 GSM 系统为主的欧洲厂商,日本公司也或多或少参与其中,历史上包括欧美的爱立信、阿尔卡特、诺基亚、朗讯、北电,以及日本的 NTT、富士通、夏普等厂商。该标准提出了 GSM(2G)-GPRS-EDGE-WCDMA(3G)的演进策略。这套系统架设在 GSM 网络上,对于系统提供商而言可以较轻松地进行过渡。2000 年左右,GSM 在亚洲十分普及,因此对 WCDMA 这套新技术的接受度也相当高。

2) CDMA 2000

CDMA 2000 是由窄带 CDMA(CDMA IS95)技术发展而来的宽带 CDMA 技术,也称为 CDMA Multi-Carrier,它是由美国高通北美公司主导提出的,摩托罗拉、Lucent 和后来加入的韩国三星都曾参与其中。而韩国则成为该标准的现行主导者。这套系统是从窄频 CDMAOne 数字标准衍生出来的,可以从原有的 CDMAOne 结构直接升级到 3G,建设成本低廉。但是,囿于使用 CDMA 的区域较少,因此 CDMA 2000 的支持者不如 CDMA 多。不过,CDMA 2000 的研发技术却是当时各标准中进度最快的。该标准提出了 CDMA IS95(2G)-CDMA 20001x-CDMA 20003x(3G)的演进策略。CDMA 20001x 被称为 2.5 代移动通信技术。CDMA 20003x 与 CDMA 20001x 的主要区别在于 CDMA 20003x 应用了多路载波技术,通过采用三载波使带宽提高。

3) TD-SCDMA

TD-SCDMA 全称为 Time Division-Synchronous CDMA(时分同步 CDMA),该标准是由中国独自制定的 3G 标准。1999 年 6 月 29 日,中国原邮电部电信科学技术研究院(大唐电信)向 ITU 提出该标准,但该技术的发明却始于西门子公司。TD-SCDMA 具有辐射低的特点,被誉为绿色 3G。该标准将智能无线、同步 CDMA 和软件无线电等当今国际领先技术融于其中,在频谱利用率、对业务支持的灵活性、频率灵活性及成本等方面都具有独特优势。另外,由于中国庞大的市场,该标准受到各大主要电信设备厂商的重视。全球一半以上的设备厂商都宣布支持 TD-SCDMA 标准。该标准曾提出"不经过 2.5 代的中间环节,直接向 3G 过渡"的口号,非常适用于当时 GSM 系统向 3G 的升级。

4）WiMAX

WiMAX 的全名是微波存取全球互通（Worldwide Interoperability for Microwave Access），又称为 802.16 无线城域网，是一种为企业和家庭用户提供"最后一英里"的宽带无线连接方案。将此技术与需要授权或免授权的微波设备相结合的成本较低，因此会扩大宽带无线市场，进而改善企业与服务供应商对该技术的认知度。2007 年 10 月 19 日，国际电信联盟在日内瓦举行的无线通信全体会议上，经过多数国家投票通过，WiMAX 正式被批准成为继 WCDMA、CDMA 2000 和 TD-SCDMA 之后的第 4 个全球 3G 标准。

2. 4G 移动通信

对移动通信技术发展史进行分析，不难发现其中存在的一些规律。例如，所有的技术都不会凭空出现，除了基于研究人员的深入研发之外，人们日益增加的服务需求也是促进技术发展的最主要动力之一。从 2G 到 3G，移动通信技术的更新速率呈加速度发展的态势。目前，业界、学界已经把 4G 移动通信技术的研究和应用发展得如日中天。

在 4G 发展的道路上，不得不提及 LTE（Long Term Evolution，长期演进）。2004 年，3GPP 在多伦多会议上首次提出了 LTE 的概念，但是它并非人们所理解的 4G 技术，而是一种 3G 与 4G 技术之间的过渡技术，俗称为 3.9G 的全球标准。它采用 OFDM 和 MIMO 作为其无线网络演进的唯一标准，改进并增强了 3G 的空中接入技术。

2008 年 6 月，3GPP 在对 LTE 进行后续研发的基础上，提出并完成了 LTE-A（LTE-Advanced 的简称）的技术需求报告，确定了 LTE-A 的最小需求：下行峰值速率 1Gb/s，上行峰值速率 500Mb/s，上下行峰值频谱利用率分别达到 15Mb/s/Hz 和 30Mb/s/Hz。与 ITU 所提供的最小技术需求指标相比较，具有非常明显的优势。换句话说，LTE-A 技术才可以称为真正的 4G 移动通信技术标准。

4G 移动通信系统的网络结构可分为三层：物理网络层、中间环境层和应用网络层。4G 移动通信对加速增长的宽带无线连接的要求提供技术上的回应，对跨越公众的和专用的、室内和室外的多种无线系统和网络提供无缝的服务。移动通信不断向数据化、高速化、宽带化、频段更高化的方向发展。移动数据、移动 IP 已经成为未来移动网的主流业务。

3. 5G 移动通信

5G 网络（5G Network）的含义为第五代移动通信网络，它的速度远超 4G 网络，大约为

4G 网络速度的 10 倍左右。从理论上讲,5G 传输速度的峰值可达到每 8 秒 1GB。如果一个视频节目的数据量为 1GB,那么,在 5G 网络平台下则只需要 8 秒即可下载完毕。5G 的到来意味着智能终端将跨越带宽的限制,在极速网络下使网络用户任性享受超高画质、3D 样式的视频或游戏等多媒体内容。

当前,5G 网络仍属于新生事物,且尚未普及,相关的技术与标准仍在探索与争论中。5G 网络通信技术传输速度之快、传输性能之稳定、传输技术之高都是其被专家视为未来网络趋势的重要属性。然而,5G 网络也同样面临着各种安全问题。例如,虚拟网络技术的脆弱性、计算存储技术与设备更新的困境,以及网络商务安全等都是亟待解决与攻克的难题。因此,只有成功规避 5G 网络的安全问题,顺应技术与市场的发展规律,才能打造理想状态的 5G 网络,使其迅速成为未来网络的主体。

相信,随着 5G 移动通信技术的不断发展,未来的人们将可以享受到更为便捷、高效、安全的应用服务。在 5G 时代,移动通信的终端设备也将发生巨大的变化。以高质量多媒体通信、低通信资费、丰富的增值服务、高智能化、多平台运行为特征的 5G 移动通信必将成为移动网络多媒体发展的新航标。

1.2.3　App 的技术特点

在智能手机系统中通常有两种应用:基于本地应用的本地应用程序和基于高端浏览器的 Web 应用程序。本机 App 位于应用程序的平台层之上,它的访问和兼容能力强,可以支持用户的线上或离线状态,消息发送和访问本地资源,及时获得照片、摄像头、拨号语音、视频等功能。它可以为用户提供最好的服务、最好的界面、最舒适方便的交互,为不同的平台提供不同的用户体验,也节省带宽成本。Web 应用程序是一个基于 Web 系统的应用,可以很容易地实现跨平台操作,用户不需要下载,可以像网站一样动态升级,也可以像 Web 一样链接到其他网站,从一个 Web 应用跳到另一个 Web 应用,成本相对较低。混合 App 是一种应用于 Web 应用与原生应用之间的应用,具有本地应用和 Web 应用的联合优势,是未来发展的主流。

1.2.4　Android 开发工具介绍

作为 App 最为流行的开发环境,Android 拥有多样化、多平台式的开发工具,编程更为便捷且容易操作。在市面上应用较为广泛的 Android 开发工具有 Eclipse、Android Studio、

Basic4android、Gimbal context ware、Titanium SDK 等,而目前在业界最为流行的是 Android Studio。

1. Eclipse

Eclipse 最初是由 IBM 公司为下一代 Java IDE 开发环境的商业软件所开发的。2001 年 11 月,Eclipse 现身于开放源码社区,并为 App 开发做出贡献,现在它由非营利软件供应商联盟 Eclipse 基金会(Eclipse Foundation)管理。在 2003 年,Eclipse 3.0 选择了 OSGi 服务平台规范作为运行时的体系结构。2007 年 6 月,稳定版 3.3 发布;2008 年 6 月,Ganymede 3.4 版发布;2009 年 6 月,Galileo 3.5 版发布;2010 年 6 月,Helios 3.6 版发布;2011 年 6 月,Indigo 3.7 版发布;2012 年 6 月,Luna 4.2 版发布;2013 年 6 月,Kepler 4.3 版发布;2014 年 6 月,Luna 4.4 版发布;2015 年 6 月,Mars 4.5 版发布。

2. Android Studio

2013 年 5 月 16 日,在 I/O 大会上,谷歌推出了一个新的 Android 开发环境——Android Studio。利用 Android Studio,开发人员能够看到应用程序同一时间在不同尺寸屏幕上呈现的效果。Android Studio 通过对控制台的改进,增加了 6 个新功能,包括优化技巧、应用翻译服务、推荐跟踪、收益图、使用版本测试和周期特性。

3. Basic4android

Basic4android 平台是一个简单但功能强大的可视化工具,是可以基于云计算的 Android 应用开发快速工具,可以在企业移动应用开发的代码库中获得所需程序。Basic4android 可用于开发和测试数据库通信,甚至可以用于开发 2D 实时游戏。Basic4android 与 Google 的 AdMob 数据库兼容,开发者可以在其应用中嵌入广告从而获得收入。

4. Gimbal context ware

Gimbal context ware 是高通实验室推出的面向 iOS 和 Android 平台的 SDK(Software Development Kit,软件开发工具包)。Gimbal 的内置函数库可以为开发人员提供特定的位置或地理隔离解决方案。Gimbal environmental awareness SDK 帮助开发者为手机用户提供及时和个性化的内容,开发者可以从 SDK 中选择加入应用程序。

5. Titanium SDK

Titanium SDK 是跨平台原生移动开发的 API，它致力于为开发者提供更高级的开发功能，开发者可以从用户界面组件或套接字接口访问系统，进而使用系统集成的原生特性和功能。Titanium 的用途是减少与纯原生应用之间的功能差异。目前，Titanium 支持 iOS、Android、黑莓和 Windows Phone。此外，Titanium 使用一个统一的 JavaScript API 来实现对特定平台特性和本机性能的代码重用，大大减少了开发人员的工作时间。

6. Vuforia

Vuforia 是一个增强现实平台，它将现实世界的物体转换成交互式体验。该平台旨在帮助开发人员创建与虚拟对象交互的真实世界对象。它利用家庭计算技术实时识别和跟踪图形图像和简单的 3D 图像，使开发人员能够跨越真实世界和数字体验之间的鸿沟。Vuforia 通过 Unity 游戏引擎为 C、Java、objective-c 和 .NET 语言提供了应用程序的编程接口。因此，Vuforia SDK 支持 iOS 和 Android 原生开发，这使得开发者在开发 Unity 引擎中的 AR 应用程序时很容易迁移到 iOS 和 Android 平台。

1.3 项目运行

1.3.1 App 分类及应用

在数字化生活的今天，人们的工作、学习、生活都离不开手机，尤其是在移动互联网如此普及的中国，手机成了集消费、娱乐、学习等多功能于一体的"百宝箱"，各种类型的 App 随时帮助我们更便捷、更有趣地生活。App 基于不同的手机操作系统有不同版本，如苹果 iOS 系统兼容的 ipa、pxl、deb 格式，谷歌 Android 系统的 APK 格式等。目前关注度最高的 App 商店有 Apple 的 App Store、Android 的 Google Play Store 和 Blackberry 的 BlackBerry App World。根据《互联网周刊》发布的 2019 年 App 分类榜单，其中包括 15 大领域的 66 个分类，如新闻资讯 App、视频 App、聊天社交 App、音乐 App、生活服务 App 等多个类目，并且每半年该周刊都会根据这些分类对当下的 App 进行排名，如表 1-1 所示。

表 1-1 App 分类目录

App应用领域	具体分类	App应用领域	具体分类	App应用领域	具体分类	App应用领域	具体分类
创新	创新	工具	使用小工具	聊天社交	社交	生活服务	外卖
	AR\VR		地图导航		社区		招聘
视频	短视频		交通票务		婚恋		租房
	在线视频		懒人应用				教育
	直播互动	电商导购	导购	新闻阅读	新闻咨询		居家
音乐	音乐		母婴电商		阅读类		医疗
	电台		跨境电商		体育		餐饮
	K歌		生鲜		汽车		家装
图像	图像美化		女性导购	企业服务	办公		家居
	图片分享		零售		电话通信		心理咨询
	相机		二手车		移动CRM		家电维修
金融理财	银行	系统工具	输入法		企业协同工作		新生活方式
	互联网金融		浏览器		企业社交		旅游综合服务
	财经		安全优化		财务软件		货运搬家
	记账		应用市场		物流信息平台		城市出行
女性母婴	女性	运动健康	运动		B2B电商		学生必备
	母婴						小清新
	美妆		健康(包括减肥)	互联网+	商超+百货		二次元
	儿童教育						

伴随着移动互联网的普及与创新应用,中国的社会化媒体格局也在发生着变化。2018年,由 Kantar Media CIC(中国领先的社会化商业资讯提供商)发布的中国社会化媒体格局概览较为准确地提炼出"复合媒体""世代""圈层营销"、Social GRP 等多个新鲜概念,勾勒出当今的中国社会化媒体生态格局,如图 1-5 所示。在新的生态格局下,手机 App 的应用不曾缺席,甚至在其中占据十分重要的位置,并以其所拥有的内容、有吸引力的 KOL、有趣的互动、跨平台推送或投放完善着整个社会化媒体的生态系统。

1.3.2 App 的发展趋势

1. 功能逐渐趋向于整合共生

在融媒体时代,除了媒体的融合共生,其功能的融合也对 App 的发展起到了至关重要的作用。当下的流行趋势即为超级应用覆盖人们的日常生活。以支付宝为例,其与淘宝功能的融合使得人们可以在支付宝上实现淘宝的大部分功能。与此同时,支付宝还专门预留

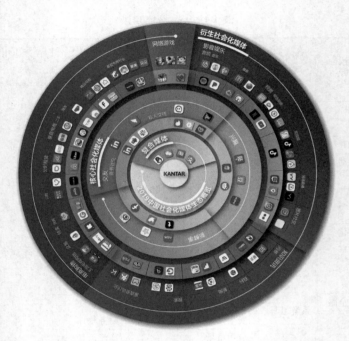

图1-5 2019年中国社会化媒体格局概览

了其他应用的入口,例如出租车、手机充值、购买电影票等服务。面对这样的发展趋势,越来越多的应用程序将通过超链接或功能模块等方式嵌入普通手机App中。

2. 设计理念逐渐趋向于扁平化

扁平化设计(Flat Design)的核心概念为去除冗余、厚重和繁杂的装饰效果。其最大的优势在于简约、符号化和适应性好。扁平化设计在App上的应用主要体现在尽可能减少按钮和选项,使得UI界面在功能上更易理解和操作,在视觉上更加干净和整齐,尽可能降低使用者的认知阻碍。这种设计理念已被大量App争相模仿与使用,最为典型的案例是将苹果公司的设计元素移植到App的界面设计中。应该说,扁平化设计对于体量小、应用性相对具体的移动App而言是相得益彰的,简洁美观的视觉体验备受开发者与用户青睐。

3. 操作方法逐渐借助于辅助类小型动画

智能手机多为大屏幕,没有按键。当用户第一次使用它们时,可能会不知所措。在这一点上,辅助动画可以快速地帮助人们理解使用方法。在当前应用程序中,对于辅助性动画的应用越来越普遍。事实证明,简明的动画对于帮助用户完成任务是很好的方法,仅仅需要一个动画示意就能够为用户提供操作上的提示,使得用户快速领会使用要领。

4. 设计观念逐渐趋于人性化

人性化设计是一种潮流,也是一种理念。一般而言,人性化设计是指在设计过程中,根据人的行为习惯、人体的生理结构、人的心理情况与思维方式等,在原有设计基本功能和性能的基础上,对产品进行优化,使体验者浏览和使用起来非常方便、舒适。对手机的设计渗入人性化思考,其实是对人的心理、生理需求和精神追求的满足,是设计中的人文关怀,是对人性的尊重。因此,App 设计师应在确保功能展现的同时,使人们在进行深度人机体验时,感受对人性的关怀。

综上所述,移动 App 的应用大大提高了我们生活、学习和工作的便利性,并能够实现高效、简洁和高度适用性。在互联网飞速发展的今天,移动 App 的应用和发展有着广阔的前景,也对我们的生活产生了潜移默化的影响。在移动 App 发展的过程中,无论是功能的整合化、设计理念的扁平化,还是操作方法的动画性、设计观念的人性化都在无形地影响着 App 的发展,最终提高整个社会的数字化水平。

1.4 项目结案

通过本项目的介绍,大家对 App 的概念、手机软硬件系统的发展、手机的关键技术、App 的技术特点、Android 开发工具、App 分类及应用以及 App 的发展趋势都有了一个较为明晰的认识。这些理论知识对我们后面进行 App 的开发具有十分重要的指导作用,它会指引我们在技术开发的时候具有对市场更为理性的认知,逐步提升对设计理念的感性运用。

1.5 项目练习

1. 手机的软件系统是如何发展的?都经历了哪几个重要时期?
2. 移动 App 是如何分类的?各举一个例子进行说明。
3. 在开发 App 时,设计理念起到什么样的作用?请举例说明。

项目 2

搭建Android开发环境

 2.1 项目目标：搭建设计 App 的 Android 开发环境

在移动数字生活时代的当下，人们对智能手机第三方应用程序（App，即 Application 的简写）的需求不断升级，而催生其发展的各种系统平台则各显其能，进行跨越式的革命。基于不同的系统，用户下载和使用的 App 的格式也有所不同。例如苹果的 iOS 系统上兼容的格式有 IPA、PXL、DEB，Android 系统上的格式有 APK，诺基亚的 S60 系统上的格式有 SIS、SISX、JAR，微软的 Windows Phone 7、Windows Phone 8 系统上的格式为 XAP，BlackBerry 平台上的格式为 ZIP 等。

Android 是嵌入式/移动应用开发平台的王者，它具有跨平台、运算速度快、体积小等特征，由其编写的应用程序可以轻松转化为各类系统平台上兼容的格式。不仅如此，具有 Java 编程经验的程序员占领了移动开发中绝大部分的市场比例，并且，具备 Java 编程能力就相当于拥有了一把万能钥匙，十分有助于理解其他平台的程序编写。

本项目的项目目标为搭建设计 App 的 Android 开发环境，请大家跟随项目的流程与具体执行步骤搭建开发环境，为 App 的设计与调试提供一个运行顺畅、界面友好的软件平台。

2.2 项目准备

2.2.1 Android 的体系介绍

Android 一词最早出现于法国作家利尔亚当（Auguste Villiers de l'Isle-Adam）在 1886 年发表的科幻小说《未来夏娃》（*L'ève future*）中。他将外表像人的机器起名为 Android。2007 年 11 月 5 日，Google 宣布 Android 为基于 Linux 平台的开源手机操作系统的名称，该平台由操作系统、中间件、用户界面和应用软件组成。

Android 的系统架构和其操作系统一样，采用了分层的架构。从系统的架构来看，Android 分为 4 层，从高层到低层分别是应用程序层、应用程序框架层、系统运行库层和 Linux 内核层，如图 2-1 所示。

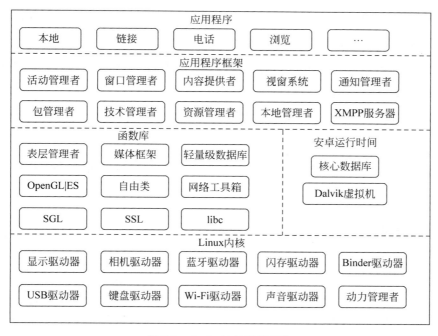

图 2-1　Android 的系统架构

1. 应用程序

Android 会和一系列核心应用程序包一起发布,该应用程序包包括客户端、SMS 短消息程序、日历、地图、浏览器、联系人管理程序等。所有的应用程序都是使用 Java 语言编写的。

2. 应用程序框架

开发人员可以完全访问核心应用程序所使用的 API 框架。该应用程序的架构设计简化了组件的重用,任何一个应用程序都可以发布它的功能块,且任何其他的应用程序都可以使用其所发布的功能块(不过要遵循框架的安全性)。同样,该应用程序重用机制也使用户可以方便地替换程序组件。

隐藏在每个应用程序后面的是一系列的服务和系统,其中包括丰富又可扩展的视图(Views),它可以用来构建应用程序,包括列表(Lists)、网格(Grids)、文本框(Text Boxes)、按钮(Buttons),甚至还可嵌入的 Web 浏览器。

应用程序框架主要包括内容提供器、资源管理器、通知管理器、活动管理器几部分。具体功能如下。

内容提供器(Content Providers)使得一个应用程序可以访问另一个应用程序的数据(例如联系人数据库),或者共享它们自己的数据。

资源管理器(Resource Manager)提供非代码资源的访问,例如本地字符串、图形和布局文件(Layout Files)。

通知管理器(Notification Manager)使得应用程序可以在状态栏中显示自定义的提示信息。

活动管理器(Activity Manager)用来管理应用程序生命周期并提供常用的导航回退功能。

3. 系统运行库

Android 包含一些 C/C++库,这些库能被 Android 系统中不同的组件使用。它们通过 Android 应用程序框架为开发者提供服务。以下是一些核心库。

(1) 系统 C 库:一个从 BSD 继承而来的标准 C 系统函数库(libc),它是专门为基于 Embedded Linux 的设备定制的。

(2) 媒体库:基于 PacketVideo OpenCORE,该库支持多种常用音频、视频格式的回放和录制,同时支持静态图像文件。其编码格式包括 MPEG4、H. 264、MP3、AAC、AMR、

JPG、PNG。

（3）Surface Manager：对显示子系统的管理，并且为多个应用程序提供了 2D 和 3D 图层的无缝融合。

＊LibWebCore 是一个最新的 Web 浏览器引擎，支持 Android 浏览器和一个可嵌入的 Web 视图。

4．Linux 内核

Android 的核心系统服务基于 Linux 2.6 内核，例如安全性、内存管理、进程管理、网络协议栈和驱动模型等都依赖于该内核。Linux 内核同时也作为硬件和软件栈之间的抽象层，它可隐藏具体硬件细节而为上层提供统一的服务。

2.2.2 Android 的安装文件介绍

在配置 Android 环境的过程中，Android SDK（Software Development Kit，软件开发工具包）是核心的软件开发工具包，有了它才能够完成程序的编写与运行。因此，深入了解这个安装文件的内部结构与相关要点是能够正确掌握 Android 程序设计的关键。下面一起查看并分析 Android SDK 的相关内容。

1．Android SDK 的结构

在安装 Android SDK 后，就可以在安装目录中看到其结构，如图 2-2 所示。

（1）add-ons：该文件夹中包含大量的附加库，例如 Google Maps 等。

（2）build-tools：该文件夹中包含 Android 开发所需的构件工具，下载并解压后，将解压出的整个文件夹，复制或者移动到 build-tools 文件夹中即可。

图 2-2　Android SDK 的结构目录

（3）docs：该文件夹中包含 Android SDK API 的参考文档与说明文档，所有的 API 都可以在这里查到。

（4）extras：该文件夹中包含与 Android 开发的相关文件，其中最为重要的是 Android 下 support 包与 Google 下的工具和驱动。

（5）platform-tools：该文件夹中包含在各平台下可以应用的工具。

（6）platforms：该文件夹包含每个平台 SDK 的相关文件，其中包括 font（字体）、res（资源）、demo（模板）等。在 platforms 文件夹中有一个名为"android.jar"的压缩包，将其解压缩后，能够更清楚地了解 API 的结构与组织方式。

（7）samples：该文件夹中是 SDK 自带的案例示范项目文件，开发者可以根据案例来自学与查看演示效果。

（8）sources：该文件夹用来放置 API 的源代码，若将源代码与具体项目进行关联，则可以通过单击类名查看其代码的实现。

（9）system-images：该文件夹中包含所有的系统图片。

（10）temp：该文件夹用于存放系统中的临时文件。

（11）tools：该文件夹中包含 SDK 中重要的工具文件。

2. SDK 文档

大家刚开始接触 Android 时，需要有相关的帮助文档给予结构与组织上的解释和说明，其中尤为重要的就是 SDK 文档。SDK 文档存放在"C：\Program Files\Android\android-sdk-windows\docs"下的 index.html 网页中。在该页面中，以超链接的形式将各种应用与工具捆绑在一起，方便开发者查找与使用；顶端有 HOME（首页）、SDK（软件开发工具包）、Dev Guide（指南）、Reference（相关）、Videos（视频）、Blog（博客）等链接，左侧为顶端下的细分帮助目录，如图 2-3 所示。与此同时，用户也可以在页面的右上角输入需要搜索的相关内容，从而进行快速搜索。

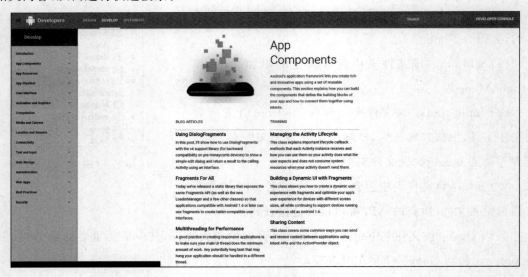

图 2-3　SDK 帮助文档索引页面

2.3 项目运行

在本书中,所有实例均通过计算机完成模拟环境的配置,并利用安装在计算机上的模拟器来演示效果及交互情况。在通常情况下,配置 Android 的开发环境分 4 步进行,如图 2-4 所示。

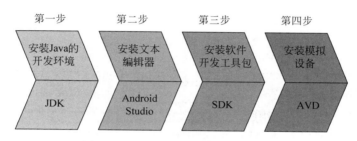

图 2-4 配置 Android 开发环境的步骤

2.3.1 安装 JDK

1. 下载 JDK

JDK(Java Development Kit)是 Sun Microsystems 针对 Java 程序开发的产品。自从 Java 推出以来,JDK 已经成为使用最广泛的 Java SDK。JDK 是整个 Java 的核心,包括 Java 运行环境、Java 工具和 Java 基础类库。学习 JDK 是学好 Java 的第一步。从 Sun 公司的 JDK 5.0 开始,提供了泛型等非常实用的功能,其版本也在不断更新,运行效率也得到了显著的提高。目前比较流行的版本为 Java SE 12。

获取 JDK 的方式为通过 Oracle 的官网下载,其网址为 http://www.oracle.com/technetwork/java/javase/downloads/index.html。如图 2-5 所示,单击图标进入下载区域,再根据所使用的操作系统与计算机位数来选择不同的下载版本,如图 2-6 所示。本书实例采用基于 Windows 64 位操作系统的 jdk-12_windows-x64_bin.exe 版本。注意,如果程序开发是基于 Linux 系统,则选择相应版本下载即可。

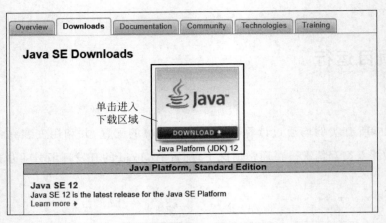

图 2-5　JDK 的下载区域

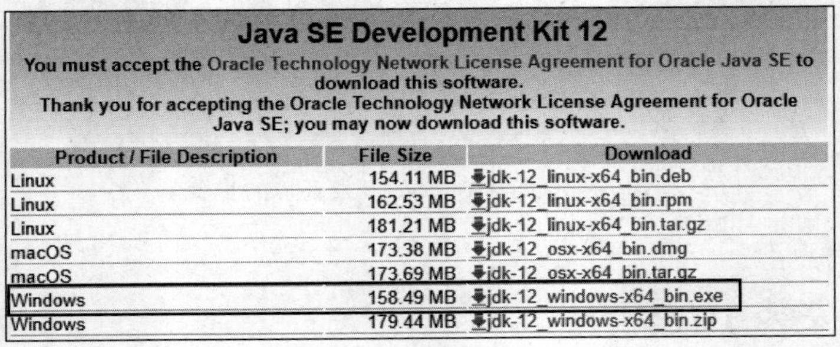
Java SE Development Kit 12

You must accept the Oracle Technology Network License Agreement for Oracle Java SE to download this software.

Thank you for accepting the Oracle Technology Network License Agreement for Oracle Java SE; you may now download this software.

Product / File Description	File Size	Download
Linux	154.11 MB	jdk-12_linux-x64_bin.deb
Linux	162.53 MB	jdk-12_linux-x64_bin.rpm
Linux	181.21 MB	jdk-12_linux-x64_bin.tar.gz
macOS	173.38 MB	jdk-12_osx-x64_bin.dmg
macOS	173.69 MB	jdk-12_osx-x64_bin.tar.gz
Windows	158.49 MB	jdk-12_windows-x64_bin.exe
Windows	179.44 MB	jdk-12_windows-x64_bin.zip

图 2-6　选择不同操作系统下的 JDK

2. 安装 JDK

下载后的 jdk-12_windows-x64_bin. exe，图标为 ，双击进入安装环节，如图 2-7 所示。

单击“下一步”按钮，进入“自定义安装”界面，可以在列表中选择要安装的功能，并指定安装位置，如图 2-8 至图 2-10 所示。注意，大家要牢记当前的安装位置，以便完成后面对于环境变量的设置。

3. 环境变量的设置

JDK 是 Java 的运行环境，没有安装 JDK 便无法测试 Java 程序。为了保证后续安装程序编辑器及相关插件时的运行效果，避免由于没有自带 JDK 的安装包而缺少运行环境，建议使用者在这一步就对环境变量进行配置。

项目2 搭建Android开发环境

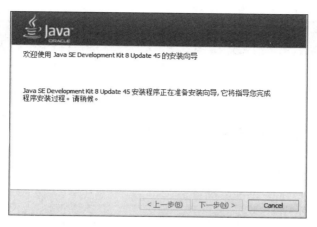

图 2-7 JDK 的安装向导界面

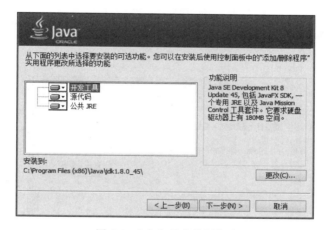

图 2-8 "自定义安装"界面

图 2-9 显示安装进度

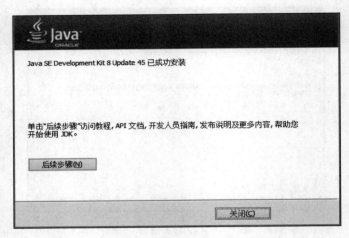

图 2-10　JDK 安装成功

右击"我的电脑",在弹出的菜单中选择"属性"命令,弹出"系统属性"对话框,如图 2-11 所示。

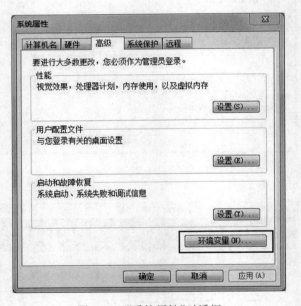

图 2-11　"系统属性"对话框

选择"高级"选项卡,单击"环境变量"按钮,出现"环境变量"对话框,如图 2-12 所示。在"系统变量"选项组中单击"新建"按钮,弹出"新建系统变量"对话框,在"变量名"文本框中输入"JAVA_HOME",如图 2-13 所示。

在"变量值"文本框中输入 JDK 的安装路径"C:\Program Files\Java\jdk-12",如图 2-13 所示。如果在"变量值"文本框中还有其他路径,不用删除,用分号隔开后再将 JDK

图 2-12 "环境变量"对话框

图 2-13 "新建系统变量"对话框

的位置输入即可。

最后,单击"确定"按钮完成环境变量的配置。

4. 运行环境的测试

在成功安装 JDK 并调整环境变量后,还需要对运行环境进行测试。

选择"开始"|"程序"|"附件"|"命令提示符"命令,打开命令提示符窗口,如图 2-14 所示。

在光标位置输入命令"java -version"。

注意:java 与 - 之间是空格;Windows 7 以上的操作系统,可以用 Win+R 键调出命令提示符窗口。

按 Enter 键后,能够看到测试效果,如果出现图 2-15 中关于 JDK 版本的文字显示,则说明安装成功。

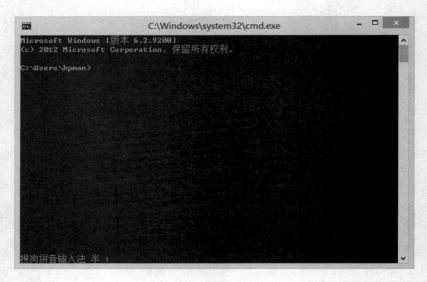

图 2-14　命令提示符窗口

图 2-15　测试效果

2.3.2　安装 Android Studio

完成第一步开发环境的配置后,就应该安装文本编辑器了。为了能够提高开发效率,简化开发流程,对于程序的编写与测试一般需要采用集成开发环境(Integrated Development Environment,IDE,可辅助开发程序的应用软件)。对于 Java 程序的开发有多款应用软件可以完成,例如早期的 Java Workshop、Borland 的 JBuilder,后期 Oracle 的 Developer 等,当下最为流行的是 Google 公司的 Android Studio。

在 Google 2013 年 I/O 大会上,Android Studio 这款开发工具被首次公布,它的出现是为了方便开发者基于 Android 系统进行开发。首先,Android Studio 解决了多分辨率的问题。Android 设备拥有大量尺寸和分辨率的屏幕,依托 Android Studio 平台,开发者可以很方便地调整在各个分辨率设备上的应用。同时，Android Studio 还解决了语言问题，多语言版本(但是没有中文版本)、支持翻译都让开发者更适应全球开发环境。Android Studio 还提供收入记录功能与 Beta Testing 的功能。

1. 下载 Android Studio

输入网址“http：//www. android-studio. org/index. php”，根据个人计算机的位数，选择单击 Windows(32-bit)或 Windows(64-bit)进入下载页面，如图 2-16 所示。之后检测到 Android Studio 的最新版本为“3. 2”，进行下载。

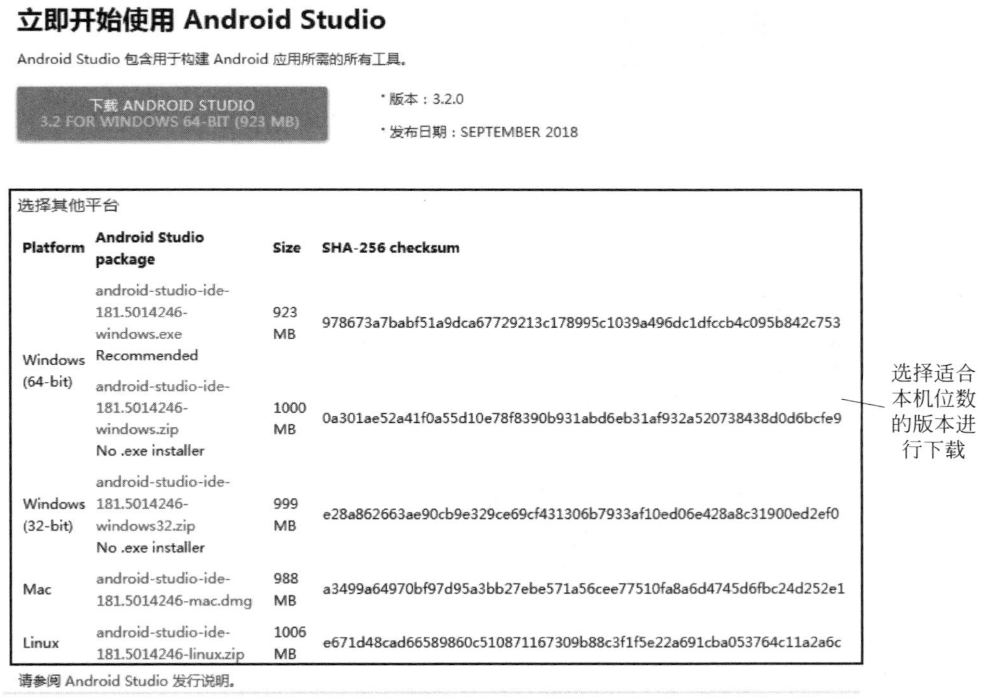
立即开始使用 Android Studio

Android Studio 包含用于构建 Android 应用所需的所有工具。

下载 ANDROID STUDIO
3.2 FOR WINDOWS 64-BIT (923 MB)

· 版本：3.2.0

· 发布日期：SEPTEMBER 2018

选择其他平台

Platform	Android Studio package	Size	SHA-256 checksum
Windows (64-bit)	android-studio-ide-181.5014246-windows.exe Recommended	923 MB	978673a7babf51a9dca67729213c178995c1039a496dc1dfccb4c095b842c753
	android-studio-ide-181.5014246-windows.zip No .exe installer	1000 MB	0a301ae52a41f0a55d10e78f8390b931abd6eb31af932a520738438d0d6bcfe9
Windows (32-bit)	android-studio-ide-181.5014246-windows32.zip No .exe installer	999 MB	e28a862663ae90cb9e329ce69cf431306b7933af10ed06e428a8c31900ed2ef0
Mac	android-studio-ide-181.5014246-mac.dmg	988 MB	a3499a64970bf97d95a3bb27ebe571a56cee77510fa8a6d4745d6fbc24d252e1
Linux	android-studio-ide-181.5014246-linux.zip	1006 MB	e671d48cad66589860c510871167309b88c3f1f5e22a691cba053764c11a2a6c

请参阅 Android Studio 发行说明。

图 2-16　Android Studio 的版本选择页面

2. 安装 Android Studio

下载 Android Studio 后，可直接双击安装文件运行软件，进入运行页面，如图 2-17 所示。

单击 Next 按钮，进入 Choose Components 界面，其中 Android Studio 为必选项，Android SDK 和 Android Virtual Device 两项建议勾选，如图 2-18 所示。

图 2-17　Android Studio 运行界面

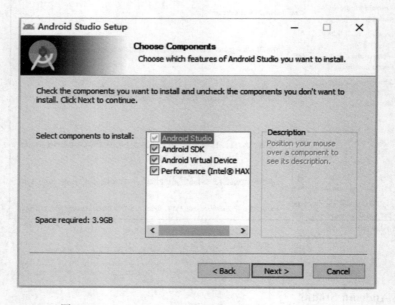

图 2-18　Android Studio 的 Choose Components 界面

单击 Next 按钮，进入 License Agreement 界面，如图 2-19 所示。然后单击 I Agree 按钮，进入 Configuration Settings 界面，为 Android Studio 和 Android SDK 选择安装目录。在默认情况下，系统会自动为其选择 C 盘指定位置来安装。如果希望安装到其他空间或更大的磁盘中，则自行指定安装目录即可，如图 2-20 所示。

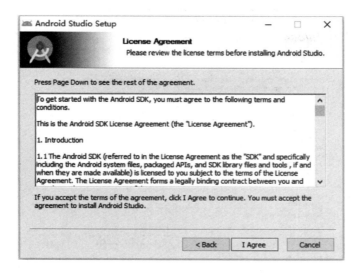

图 2-19　License Agreement 界面

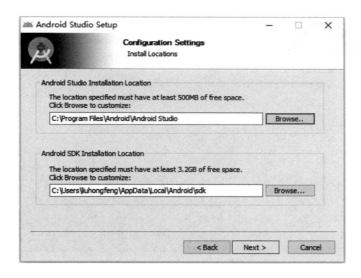

图 2-20　Configuration Settings 界面

单击 Next 按钮，进入 Choose Start Menu Folder 界面，为 Android Studio 设置"开始"菜单文件夹的名字，一般会自动默认为 Android Studio，无须修改，直接单击 Install 按钮安装即可，如图 2-21 所示。

进入 Installing 界面，Android Studio 开始安装，并以进度条的形式显示，如图 2-22 所示。

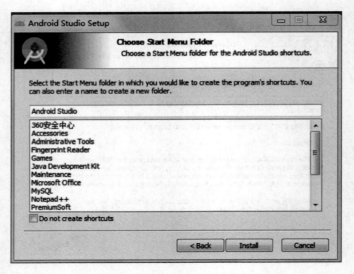

图 2-21　Choose Start Menu Folder 界面

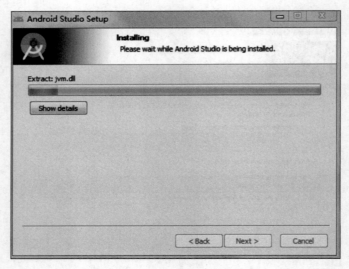

图 2-22　Installing 界面

在 Android Studio 安装完成后，单击 Next 按钮，进入 Completing Android Studio Setup 界面，单击 Finish 按钮，完成 Android Studio 的安装，并开启 Android Studio，如图 2-23 所示。

2.3.3　配置 Android SDK

在正式启动 Android Studio 之前，建议大家为 Android Studio 配置 SDK。具体的配置

图 2-23　Completing Android Studio Setup 界面

方法十分简单,即找到 Android Studio 的安装目录,打开 bin 文件夹下的 idea.properties 文件,如图 2-24 所示,采用文本编辑工具(例如记事本、写字板等)打开该文件,并在文本的最后添加一句代码(如图 2-25 所示)。

```
disable.android.first.run = true
```

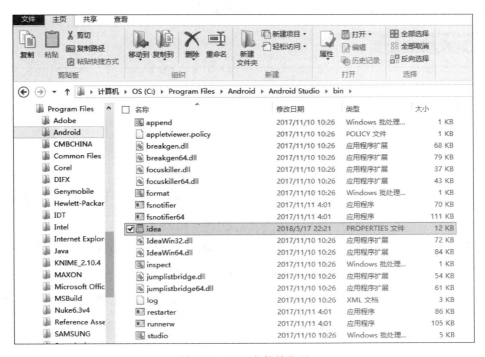

图 2-24　idea 文件的位置

注意：在对 idea.properties 文件进行重新编辑时，可能会由于管理员权限的问题而无法保存新文件。这时将修改后的文件复制、粘贴到其他目录下，然后将原来的文件删除，再将新文件拖回到 bin 文件夹下即可。

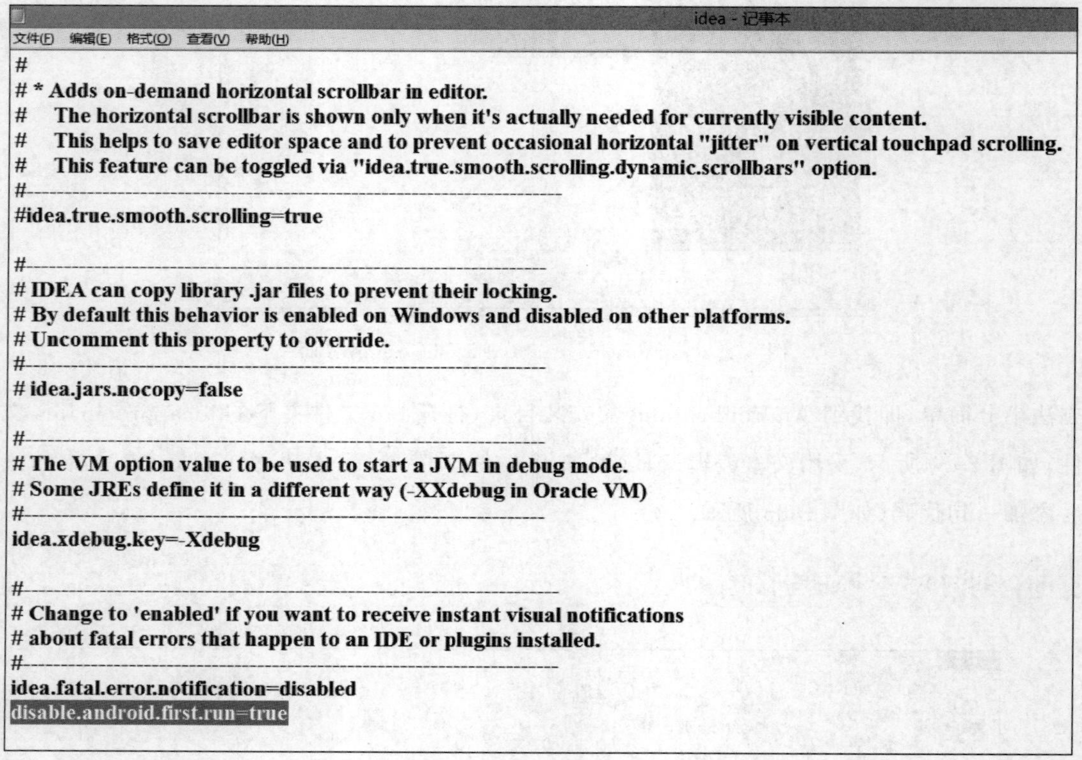

```
#
# * Adds on-demand horizontal scrollbar in editor.
#   The horizontal scrollbar is shown only when it's actually needed for currently visible content.
#   This helps to save editor space and to prevent occasional horizontal "jitter" on vertical touchpad scrolling.
#   This feature can be toggled via "idea.true.smooth.scrolling.dynamic.scrollbars" option.
#-------------------------------------------------------------
#idea.true.smooth.scrolling=true

#-------------------------------------------------------------
# IDEA can copy library .jar files to prevent their locking.
# By default this behavior is enabled on Windows and disabled on other platforms.
# Uncomment this property to override.
#-------------------------------------------------------------
# idea.jars.nocopy=false

#-------------------------------------------------------------
# The VM option value to be used to start a JVM in debug mode.
# Some JREs define it in a different way (-XXdebug in Oracle VM)
#-------------------------------------------------------------
idea.xdebug.key=-Xdebug

#-------------------------------------------------------------
# Change to 'enabled' if you want to receive instant visual notifications
# about fatal errors that happen to an IDE or plugins installed.
#-------------------------------------------------------------
idea.fatal.error.notification=disabled
disable.android.first.run=true
```

图 2-25 添加代码

2.3.4 安装 AVD

在使用 Android Studio 进行程序编写时，需要用模拟器来显示程序效果，这就需要为 Android 配置安卓虚拟设备，即 AVD(Android Virtual Device)。

首先启动 Android Studio，由于是第一次启动，所以会出现 Android Studio 开启方式的选项窗口，单击 Start a new Android Studio project 选项，如图 2-26 所示，进入 Android Studio 的主界面。以上步骤操作一次后，再启动 Android Studio 时，便会自动进入上一次编辑的 Android 项目中。

图 2-26　Android Studio 开启方式的选项窗口

然后，为了创建 AVD，单击 Android Studio 工具栏中的 AVD Manager 按钮，如图 2-27 所示，并在弹出的 Android Virtual Device Manager 窗口中单击＋Create Virtual Device 按钮，如图 2-28 所示。随即弹出 Virtual Device Configuration 窗口，可以选择设备定义，包括设备类型、型号、尺寸、分辨率、像素等信息，如图 2-29 所示。

图 2-27　单击 AVD Manager 按钮

在选定设备后单击 Next 按钮，进入 System Image 界面，选择一个 API Level。如果它并未安装过，则会出现 null Download 字样，单击 Download 按钮，便会弹出 API 的安装许可界面，单击 Accept 按钮确定快速安装 SDK，再单击 Finish 按钮完成 API 组件的安装。

在组件安装完成后，再次进入 Android Virtual Device Manager 界面，能够看到已经配置成功的模拟器。单击 ■ 按钮运行该虚拟设备即可，如图 2-30 所示。

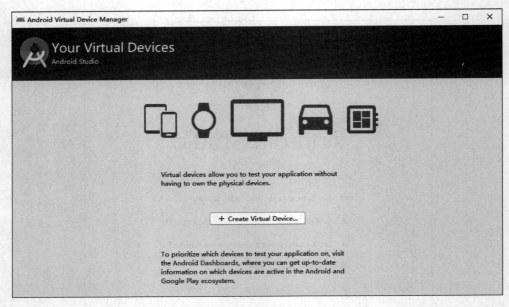

图 2-28　单击+Create Virtual Device 按钮

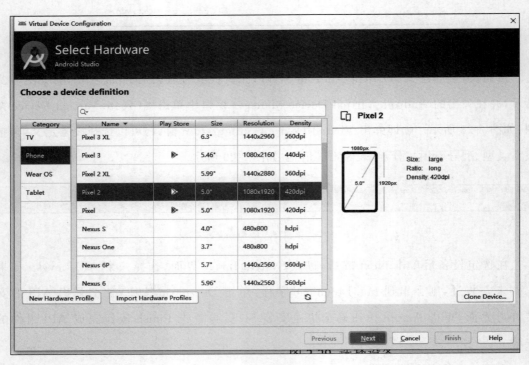

图 2-29　选择设备

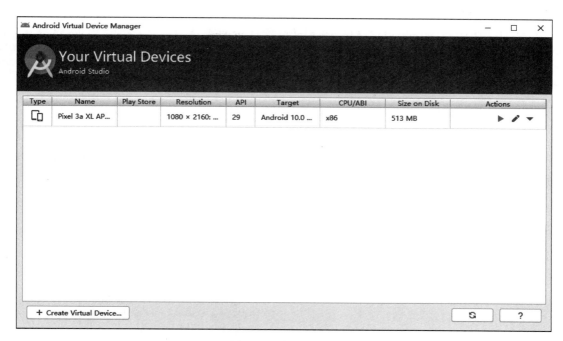

图 2-30　我的虚拟器

2.3.5　Android Studio 操作指南

1. Android Studio 的主菜单栏

Android Studio 在不同的操作系统下有不同版本的安装程序，但是其菜单栏的设置和主要功能基本相同。这里以常用的 Windows 操作系统为例，了解一下 Android Studio 的主菜单栏选项。

Android Studio 的主菜单栏选项包括 File、Edit、View、Navigate、Code、Analyze、Refactor、Build、Run、Tools、VCS、Window 和 Help，如图 2-31 所示。Mac 版本的 Android Studio 还会出现 Android SDK Manager 菜单选项。

图 2-31　Windows 操作系统下 Android Studio 的主菜单栏选项

1）Android SDK Manager 菜单选项

在 Mac 版的 Android SDK Manager 菜单选项中包含 About Android SDK Manager（Android SDK Manager 的介绍）、Preferences（属性）、Services（服务器）、Quit Android SDK Manager（退出 Android SDK Manager）等选项，如图 2-32 所示。

2）File 菜单选项

File 菜单选项中提供了对项目文件的各种操作，包括 New（新建）、Open（打开）、Settings（设置）、Project Structure（项目结构）、Save All（保存）和 Exit（退出）等，如图 2-33 所示。

图 2-32　Android SDK Manager 菜单选项

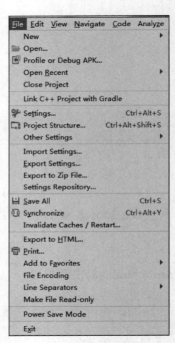

图 2-33　File 菜单选项下拉列表

在 File 菜单选项中，较为重要的选项是 Settings（设置），用户可以通过该选项对 Android Studio 进行配置。单击 File 菜单中的 Settings 选项，弹出 Settings 对话框，其中包括 Appearance & Behavior、Keymap、Editor、Plugins、Version Control、Build、Execution、Deployment、Languages & Frameworks、Tools、Kotlin Compiler 等设置功能，如图 2-34 所示。

New 选项为新建，其包括 New Project（新建项目）、Import Project（导入项目）、New Module（新建模板）、Import Module（导入模板）以及新建文件夹、新建类等多种操作，如图 2-35 所示。

项目2　搭建Android开发环境

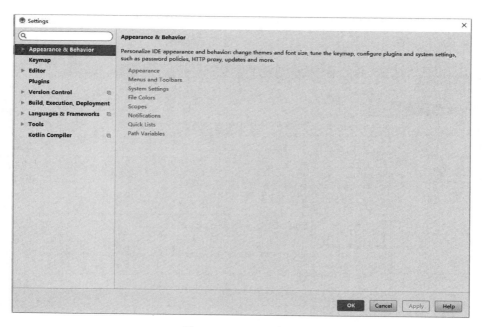

图 2-34　Settings 窗口

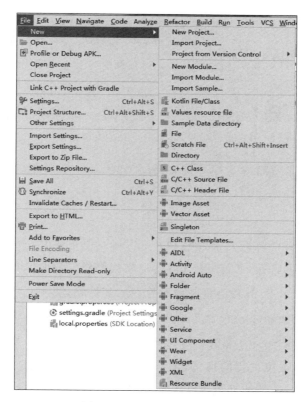

图 2-35　New 选项下拉列表

在 Project Structure 对话框中能够配置项目的 SDK、Ads、JDK、NDK 以及 Authentication、Notifications 等,如图 2-36 所示。

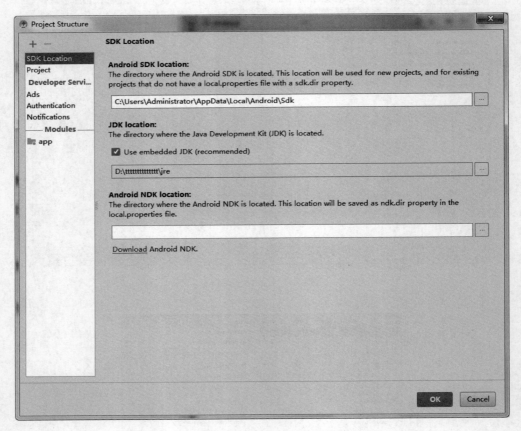

图 2-36　Project Structure 对话框

Import Settings 选项能够为用户提供导入配置文件等功能,其窗口显示如图 2-37 所示。

3) Edit 菜单选项

Edit 菜单选项为用户提供了 Undo(撤销)、Copy(复制)、Paste(粘贴)、Find(查找)、Macros(宏)、Select All(选择)、Convert Indents(转换缩进)等功能,如图 2-38 所示。其中,Copy 有 3 种方式:Copy Path 意为复制最近打开的文件路径;Copy as Plain Text 意为复制无格式文本;Copy Reference 意为复制引用。Paste 有两种方式:Paste from History 意为从历史中粘贴;Paste Simple 意为粘贴无格式文本。

Find 选项是一个比较重要的应用,其中包括 Find in Path、Replace in Path、Search Structurally、Find Usages 等搜索和选择功能,如图 2-39 所示。

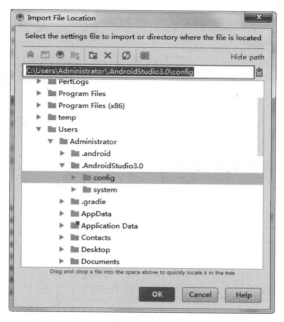

图 2-37　Import Settings 对应的窗口

图 2-38　Edit 菜单选项下拉列表

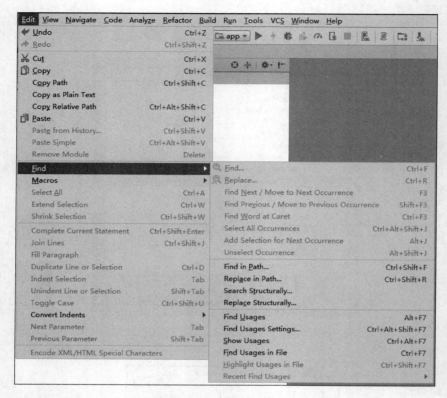

图 2-39 Find 选项下拉列表

Macros 选项为宏功能，能够帮助用户实现开启宏记录、编辑宏等操作，具体操作功能如图 2-40 所示。

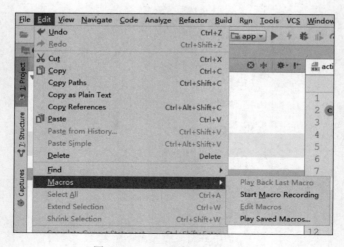

图 2-40 Macros 选项下拉列表

4) View 菜单选项

View 菜单能够对视图进行操控,可以帮助用户便捷地设置界面及窗口的显示效果,包括 Tool Windows(工具窗口)、Open Module Settings(打开视图模板设置)、Toolbar(工具栏的视图设置)、Enter Full Screen(设置全屏)等功能,如图 2-41 所示。基于这些功能,用户能够随意设置菜单栏、工具栏、导航栏及窗口的显示方式与禁用与否,这是体现个性化的功能之一。

5) Navigate 菜单选项

Navigate 意为导航,该菜单多用于结构复杂、内容丰富、文件类型多样化的项目,为用户提供便于在项目中处理多文件的操作,如图 2-42 所示。

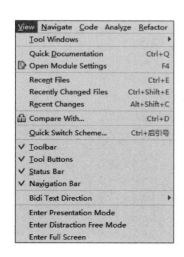

图 2-41　View 菜单选项下拉列表

图 2-42　Navigate 菜单选项下拉列表

在 Navigate 菜单选项中,Bookmarks 意为书签,用于为某行重要文件、方法、代码等做标记,如图 2-43 所示。一般而言,用户在做项目时经常会回看或反复查看某个文件或某个方法,为了便于查看这些访问率较高的文件,可以使用 Bookmarks 对其进行标注。

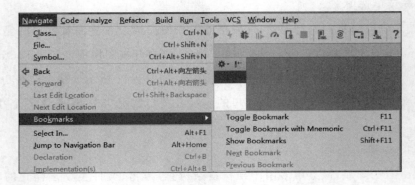

图 2-43　Bookmarks 选项下拉列表

6）Code 菜单选项

Code 菜单提供专门针对代码操作设置的选项，其中包括 Override Methods（重写方法）、Implement Methods（实施方法）、Delegate Methods（授权方法）、Reformat Code（重新格式化代码）、Rearrange Code（重新安排代码）、Move Line Down（移动代码行）、Insert Live Template（插入 Live 模板）等操作，如图 2-44 所示。值得注意的是，这些对于代码的控制一般都用于 Java 文件，非 Java 文件不可用。

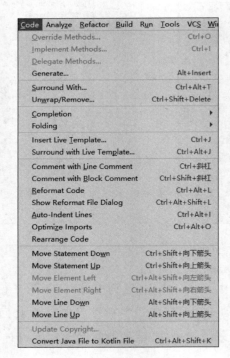

图 2-44　Code 菜单选项下拉列表

项目2　搭建Android开发环境

Code 菜单中的 Completion 选项为用户提供了多种代码的实现方式,如图 2-45 所示。

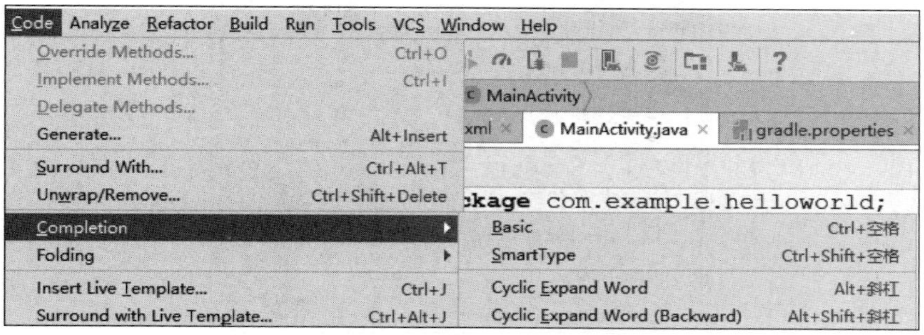

图 2-45　Completion 选项下拉列表

7) Analyze 菜单选项

Analyze 意为分析代码,该菜单为用户提供了提高代码质量和效率的多种方法,包括 Inspect Code(检查代码)、Code Cleanup(清除代码)、Analyze Dependencies(深度分析)、Analyze Backward Dependencies(向后深度分析)、Analyze Module Dependencies(模块深度分析)等操作,如图 2-46 所示。

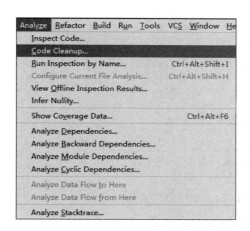

图 2-46　Analyze 菜单选项下拉列表

Inspect Code 选项对应的对话框能够为用户提供对整个项目或部分的检查,并将检查文件保存在指定位置,如图 2-47 所示。

8) Refactor 菜单选项

Refactor 菜单为用户提供了重构的各种方法与功能,具体如图 2-48 所示。

Refactor This 选项能够帮助用户对项目或其他文件进行 Rename(重命名)、Move(移动)、Copy(复制)等操作,如图 2-49 所示。

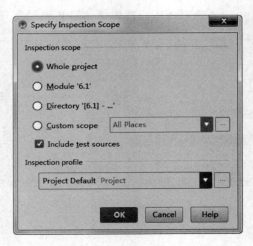

图 2-47　Inspect Code 选项对应的对话框

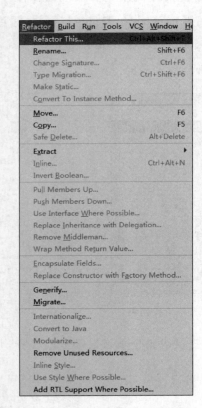

图 2-48　Refactor 菜单选项下拉列表

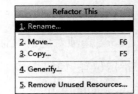

图 2-49　Refactor This 选项窗口

Extract 选项能够帮助用户抽取 Parameter Object（参数对象）、Delegate（授权）、Style（风格）、Layout（布局）等信息，如图 2-50 所示。

项目2　搭建Android开发环境

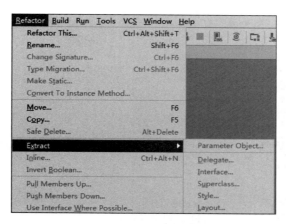

图 2-50　Extract 选项下拉列表

9) Build 菜单选项

Build 菜单为用户提供了对项目进行构建及相关配置等功能，包括 Make Project（构建项目）、Make Modules（构建模板）、Clean Project（清除项目）、Rebuild Project（重构项目）以及 Build APK(s)（构建 APK）等操作，如图 2-51 所示。

图 2-51　Build 菜单选项下拉列表

10) Run 菜单选项

Run 菜单为用户提供了用于调试和运行程序的各种功能，其中包括 Run App（运行 App）、Debug App（纠错）等操作，如图 2-52 所示。这些操作能够使用户快速地实现程序的调试和运行。

11) Tools 菜单选项

Tools 菜单为用户提供了一些与 Android 开发相关性不强的工具，包括 Tasks & Contexts（任务与环境的查看与设置）、Generate JavaDoc（生成 Java 文件）、App Links Assistant（连接辅助功能）等操作，如图 2-53 所示。

43

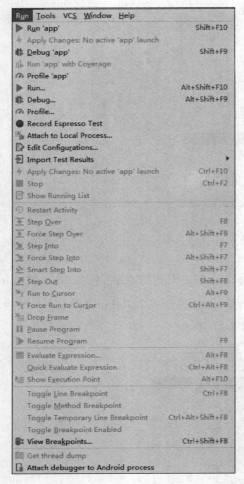
图 2-52　Run 菜单选项下拉列表

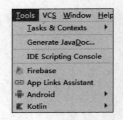
图 2-53　Tools 菜单选项下拉列表

Tasks & Contexts 选项中包括对任务的各项操作和环境的设置，例如 Switch Task（转换任务）、Open Task（打开任务）、Close Active Task（关闭活动任务）、Save Context（保存环境）、Load Context（装载环境）、Clear Context（清除环境）等操作，如图 2-54 所示。

12）VCS 菜单选项

VCS 意为版本控制系统，该菜单提供了帮助用户设置与管理版本相关问题的选项，其中包括 Local History（本地版本历史）、Enable Version Control Integration（版本控制集成）、Checkout from Version Control（检验版本控制）、Import into Version Control（导入版本控制）等，如图 2-55 所示。

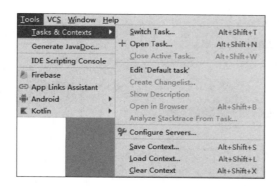

图 2-54　Tasks & Contexts 选项下拉列表

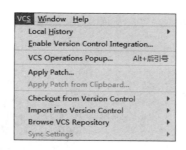

图 2-55　VCS 菜单选项下拉列表

Checkout from Version Control(检验版本控制)选项中包含各种控制版本的系统,如图 2-56 所示。其中,CVS 是 2000 年初期流行于世的版本控制系统;Subversion 是紧随其后的流行系统,它的最大优势是其事务性提交功能。目前,Git 是 Android 开发中应用最为广泛的版本控制系统。

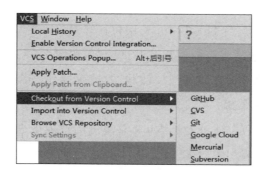

图 2-56　Checkout from Version Control 选项下拉列表

Import into Version Control 选项中为用户提供了导入版本控制系统的方法与操作,包括 CVS、Git、Subversion 等系统的导入,如图 2-57 所示。

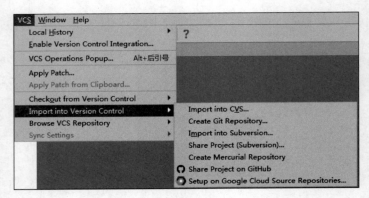

图 2-57　Import into Version Control 选项下拉列表

Browse VCS Repository(浏览 VCS 仓库)选项能够对不同版本控制系统的仓库进行浏览,如图 2-58 所示。

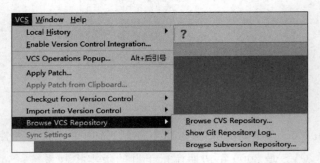

图 2-58　Browse VCS Repository 选项下拉列表

2. Android Studio 的工具栏

Android Studio 的工具栏(The Toolbar)中包括针对代码文本编辑的常用按钮,以及不同类型的管理器按钮,还包括程序的运行及调试按钮,如图 2-59 所示。对于文本编辑的常用按钮而言,包括新建、保存、剪切、复制、粘贴、撤销等操作。管理器按钮包括 SDK 管理器、Android 虚拟设备管理器等。

工具栏中的工具项目可以通过勾选 View 菜单中的选项来进行取舍,每一个工具按钮都包含相应的菜单下拉列表和快捷键方式。

3. Android Studio 的导航栏

Android Studio 的导航栏(The Navigation Bar)是集成开发环境(IDE)的基础功能。在导航栏中能够清晰地看到 Android 项目的内部结构,其中包含组成一个完整 App 项目的所

项目2 搭建Android开发环境

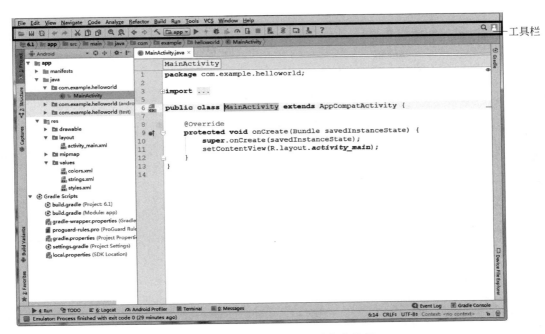

图 2-59　Android Studio 的主菜单栏

有文件,如图 2-60 所示。导航栏以水平箭头的链状结构方式来呈现,越靠近左侧越接近根目录。单击导航栏上的选项卡能够快速地选中并进入该文件内部。

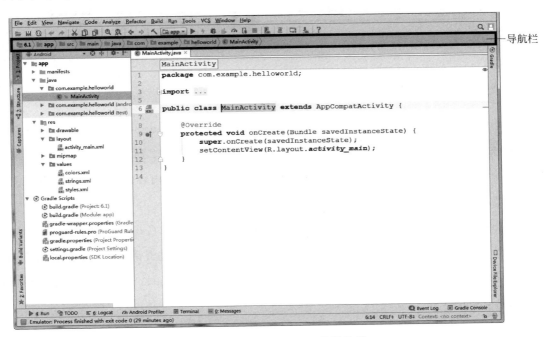

图 2-60　Android Studio 的导航栏

4. Android Studio 的状态栏

Android Studio 的状态栏(The Status Bar)用于显示当前项目或文档的状态信息,包括正在运行的进程、Git 版本本库等信息,如图 2-61 所示。

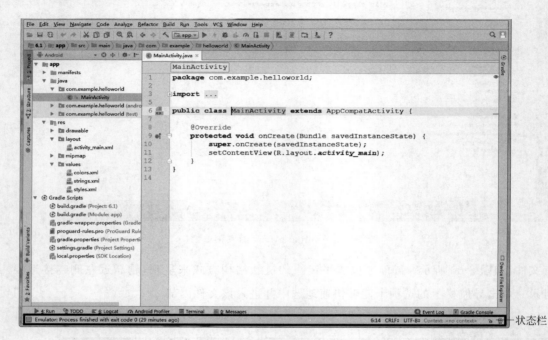

图 2-61 Android Studio 的状态栏

5. Android Studio 的编辑器

Android Studio 的编辑器(The Editor)在默认情况下位于 Android Studio 界面的中心,并以选项卡的形式将所选中的文件显示在编辑器窗口内,如图 2-62 所示。在编辑器的上部是当前打开文件的名字,当单击该文件的名称选项卡时,其文件内容便会呈现在编辑器的窗口中央。

Android Studio 的编辑器(The Editor)的左侧即为边列(The Gutter),用于传递代码信息,其鲜明的标识即在代码中以小色卡或者小图标来显示可视化资源,同时也能够用于断点的设置、代码的折叠、约定代码范围等;右侧为标记栏(The Marker Bar),用于指示资源文件中重要行的位置。

项目2　搭建Android开发环境

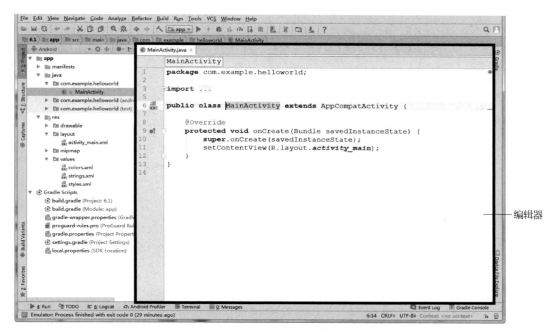

图 2-62　Android Studio 的编辑器

6．使用代码折叠

使用代码折叠（Using Code Folding）可以为编辑器节省屏幕空间，以隐藏特定代码块的方式显示代码的框架性信息，如图 2-63 所示。在代码折叠框内部包含 3 种图标，即折叠图标 ▣ 、向下箭头图标 ▽ 、向上箭头图标 △ 。

折叠图标代表某个已经被折叠了的代码块，被折叠的代码会呈现为一个省略号，且以高亮的浅绿色标注；向下箭头图标代表可以被折叠的一段代码的起始位置；向上箭头图标代表可以被折叠的一段代码的结束位置。

7．工具按钮

在 Android Studio 界面中包含多组工具按钮，这些工具按钮平时以默认的状态显示在界面上，当用户需要查看全部工具按钮时，可以通过主菜单栏中的 View 选项进行调用。

1）项目工具窗口

项目工具窗口（The Project Tool Window）位于编辑器的左侧，可以通过 3 种模式对 App 项目进行查看与调用，分别为 Android、Project 和 Package，默认为 Android 模式。在该窗口中 App 项目以树形结构展现，能够使用户十分便捷地切换文本，如图 2-64 所示。

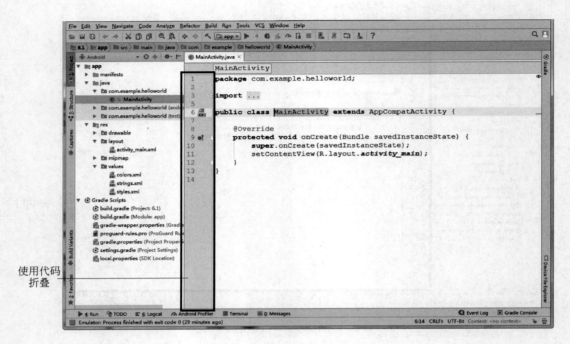

图 2-63 Android Studio 使用代码折叠

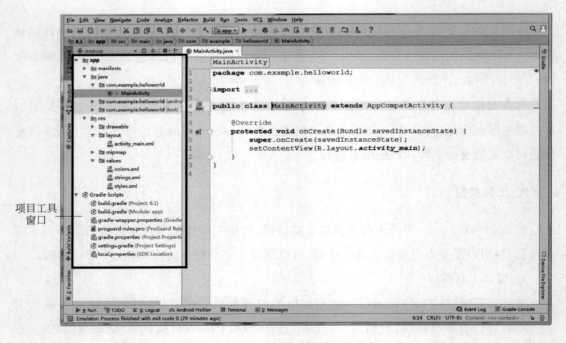

图 2-64 Android Studio 的项目工具窗口

2）结构工具窗口

结构工具窗口（The Stucture Tool Window）用来展现 App 项目文件中元素的层次结构，如图 2-65 所示，一般适用于超大源文件中的元素。

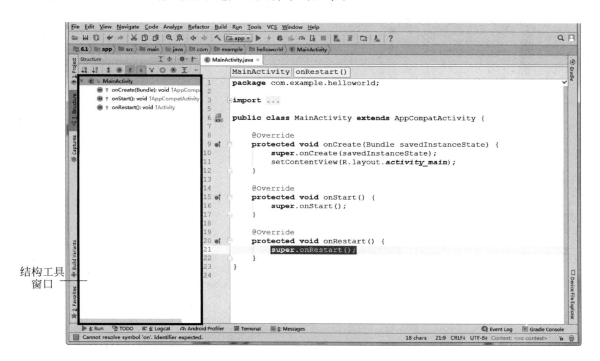

图 2-65　Android Studio 的结构工具窗口

3）收藏夹工具窗口

收藏夹工具窗口（The Favorites Tool Window）中包含逻辑分组目录、书签目录（Bookmarks）以及断点目录（Breakpoints），如图 2-66 所示。

（1）逻辑分组目录：收藏夹窗口能够将位于 App 项目中不同位置的文件进行逻辑分组。例如，用户在编辑器选项卡中打开 MainActivity.java 文件，右击该文件的选项卡名，选择 Add All to Favorites 选项，在弹出的窗口中输入该分组的名字，如"Activity"，再单击 OK 按钮退出即可。通过上述操作，在逻辑分组中就会出现一个名为"Activity"的目录。如果要将其他文件也加入这个目录，则采用上述方式将其加入"Activity"的目录中。

（2）书签目录：书签能够帮助用户快速找到特定代码行，所有的书签都会被保存在书签目录中。添加书签的快捷键为 F11，取消书签则在选中标记了书签的代码行后再次按 F11

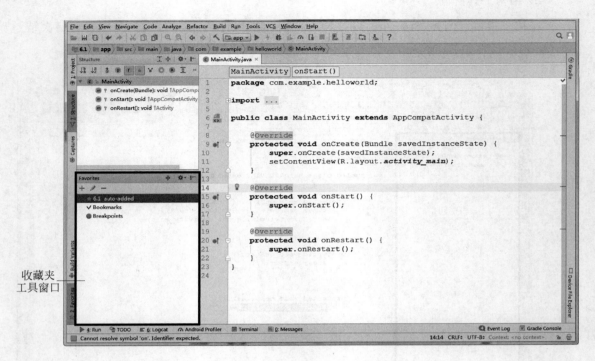

图 2-66　Android Studio 的收藏夹工具窗口

键即可。

（3）断点目录：断点是在程序调试时进行片段调试时设置的标记。在被设置了断点的代码行上会出现一个红色圆圈，且整行代码都是红色高亮的。App 项目中的所有断点设置都能够在收藏夹的断点目录中找到。

8．Android Studio 的实用快捷键

Android Studio 的实用快捷键如表 2-1 所示。

表 2-1　Android Studio 的实用快捷键

快 捷 键	实 现 功 能
Ctrl＋B	用于跳入/跳出方法或者资源文件。将鼠标光标定位到某个方法或者资源 id 的调用处，按 Ctrl＋B 键，将会跳入该方法或者资源文件内部，功能等同于按 Ctrl＋鼠标左键。如果将鼠标光标定位到方法定义处或者资源文件内部，按 Ctrl＋B 键将会返回调用处

续表

快 捷 键	实 现 功 能
Ctrl+O	用于查看父类中的方法,并可以选择父类方法进行覆盖。将鼠标光标定位到类中代码的任意位置,按 Ctrl+O 键,将会在打开的面板中查看到所有父类中的所有非私有方法,选择某个方法按 Enter 键即可覆盖父类方法
Ctrl+K	用于 SVN 提交代码
Ctrl+T	用于 SVN 更新代码
Ctrl+H	用于查看类的上下继承关系。将鼠标光标定位在类中的任何一个位置,然后按 Ctrl+H 键,将会打开一个面板,在这个面板中会依照层级显示出当前类的所有父类和子类
Ctrl+W	用于选中代码块,多次按 Ctrl+W 键将逐步扩大选择范围
Ctrl+E	用于显示最近打开的文件,可以快速地再次打开这些文件
Ctrl+U	用于快速跳转至父类,或者快速跳转到父类中的某个方法。将鼠标光标定位到类名上,按 Ctrl+U 键,将会打开当前类的父类;如果当前类有多个父类,则会提示要打开的父类。如果一个类中的方法覆盖了其父类的方法,那么将鼠标光标定位到子类的方法,按 Ctrl+U 键,将会跳转到被覆盖的父类方法中
Ctrl+G	用于显示鼠标光标当前位置在代码文件中的行/列数,可以理解为光标在代码中的横/纵坐标
Ctrl+F12	用于查看类中的所有变量、方法、内部类、内部接口。将鼠标光标定位到当前类文件的任意位置,按 Ctrl+F12 键会弹出显示类中所有变量、方法、内部类、内部接口的对话框,然后按 ↑、↓ 键可以选择某个变量、方法、内部类、内部接口,接着按 Enter 键可以快速定位到该变量、方法、内部类、内部接口
Ctrl+F11	用于在鼠标光标所在行添加书签。如果文件中的代码特别多,那么书签将是一个非常实用的功能,它可以帮助用户标记代码中的重要位置,方便下次快速定位到这些重要位置
Shift+F11	用于查看书签,可以快速查看之前标记的书签
Ctrl+Shift+F12	用于快速调整代码编辑窗口的大小
Ctrl+↑/↓	用于固定光标上下移动代码
Alt+↑/↓	用于在内部接口、内部类和方法之间跳转
Ctrl+Shift+Backspace	用于回到上一次编辑的位置
Alt+数字	用于打开相应数字的面板。例如终端控制台面板对应的数字是 6,那么按 Alt+6 键可以快速展开或关闭控制台面板

续表

快 捷 键	实 现 功 能
Ctrl+Shift+I	用于快速查看某个方法、类、接口的内容。将鼠标光标定位到某个方法、类名、接口名,然后按 Ctrl+Shift+I 键,将会在当前光标位置显示该方法、类、接口的内容
Shift+Esc	用于关闭当前打开的面板
Alt+J	用于选择多个相同名字的关键字、方法、类、接口,然后同时更改
Ctrl+Tab	用于切换面板或文件,功能类似于 Windows 下的 Alt+Tab 键。在切换面板/文件的对话框中,选中某个面板或文件,接着按 Backspace 键即可关闭该面板或文件
Ctrl+Shift+Enter	用于快速补全语句。例如 if(){}、switch(){}代码块,只要输入 if 或者 switch(甚至 sw),接着按 Ctrl+Shift+Enter 键,就可以快速完成代码块
Ctrl+Alt+M	用于快速抽取方法。选中代码块,然后按 Ctrl+Alt+M 键可以快速将选中的代码块抽取为一个方法
Ctrl+Alt+T	用于快速包裹代码块。选中一段代码,然后按 Ctrl+Alt+T 键,可以选择要对选中代码块进行的操作,例如 if/else、do/while、try/catch/finally 等
Ctrl+Alt+L	用于代码的格式化
Ctrl+N	用于快速查找类。按下 Ctrl+N 键会弹出输入类名的对话框,在该对话框的搜索框中输入要查找的类名,即可开始进行模糊检索,这样可以快速找到需要查找的类,这在类文件非常多的工程里面特别实用
Ctrl+Shift+N	用于快速查找文件。功能和 Ctrl+N 键类似,但是除了可以搜索类文件之外,还可以搜索当前工程下的所有文件,这同样是一个经常用到的特别实用的功能
Double Shift	用于全局搜索。功能和 Ctrl+N 键、Ctrl+Shift+N 键类似,但是搜索的范围更广,支持符号检索,除了 Ctrl+N 键、Ctrl+Shift+N 键的检索内容外,还可以搜索到变量、资源 id 等
Ctrl+Alt+Space	用于类名或接口名提示。输入一个不完整的类名或者接口名,按 Ctrl+Alt+Space 键,会给出完整类名或接口名的提示
Ctrl+Q	用于显示注释文档。将鼠标光标定位到某个类名、接口名或者方法名,按 Ctrl+Q 键,会显示出该类、接口、方法的注释
Ctrl+PageUp/PageDown	用于将光标定位到当前文件的第一行/最后一行
Shift+Left Click	用于关闭当前文件
Ctrl+Alt+B	用于跳转到抽象方法的实现。将鼠标光标定位到某个抽象方法,然后按 Ctrl+Alt+B 键,会快速跳转到该抽象方法的具体实现处,如果该抽象方法有多个具体实现,那么会弹出选择框让用户进行选择

续表

快 捷 键	实 现 功 能
Ctrl+Shift+U	用于快速进行大小写转换
Ctrl+Shift+Alt+S	用于打开 Project Structure 面板
Ctrl+F	用于在当前文件中搜索输入的内容
Ctrl+R	用于在当前文件中替换输入的内容
Ctrl+Shift+F	用于全局搜索
Ctrl+Shift+R	用于全局替换
Shift+F6	用于快速重命名。选中某个类、变量、资源 id 等之后,可以快速重命名,只要改动一个位置,其他地方都会自动全部重命名
Alt+F7	用于快速查找某个类、方法、变量、资源 id 被调用的地方
Ctrl+Shift+Alt+I	用于对项目进行审查。按下 Ctrl+Shift+Alt+I 键,会弹出搜索审查项的输入框,输入关键字可以检索需要审查的内容,例如输入 unused resource 即可搜索项目中没有使用到的资源文件。此外,在菜单栏中选择 Analyze|Inspect Code 命令,或者右击当前工程,选择 Analyze|Inspect Code 命令,可以对项目进行 Lint 审查
Ctrl+D	用于快速复制行
Ctrl+Shift+↑/↓	用于上下移动代码。如果是方法中的代码,只能在方法内部移动,不能跨方法
Shift+Alt+↑/↓	用于上下移动代码,可以跨方法移动
Shift+F10	用于启动 Module
Shift+F9	用于调试 Module
Ctrl+F9	用于构建 Project
Alt+Insert	用于快速插入代码,可以快速生成构造方法、Getter/Setter 方法等
Alt+Enter	用于快速修复错误

2.4 项目结案

通过本项目的学习,相信大家已经掌握了 Android Studio 环境配置的方法,在这个过程中,需要从网上下载相关的安装包,并在安装后与本机进行连接。只有保证每一个步骤都成功进行,才能在后面顺利展开建立 App 程序与调试的工作。另外,本项目还用大量的篇幅介绍了 Android Studio 的基本操作。子曰:"工欲善其事,必先利其器",Android

Studio 的开发环境就是开发 App 的重要工具,希望大家能够耐心完成前面的操作。

2.5 项目练习

1. 尝试在 Android Studio 中新建一个程序,并将该程序命名为"与 App 的第一次约会"。

2. 打开 Android SDK 的结构目录,分别查看每个文件夹,并感受里面的内容与 App 设计的关系。

3. 打开 SDK 的帮助文档索引页面,查看里面的内容,并对自己认识的类进行简单描述。

项目3

创建第一个App

3.1 项目目标：用 Android Studio 创建 App

使用 Android Studio 可以十分便捷且以全程可视化的方式完成 App 的创建、运行与调试。相较于其他 Android 开发工具，响应速度更快、UI 主题更具设计性、调试程序更加智能等优势让 Android Studio 表现出更加优秀的属性。因此，使用 Android Studio 进行 App 的创建与设计将是一件相当快乐的事情，请大家保持愉快的心情开始 App 之旅。

3.2 项目准备

3.2.1 Android 的内部结构

为了保证 Android 程序结构的一致性，Android Studio 为每一个程序设置了相同的内部结构，该结构在 Android 项目建立之初就已经存在了。程序的内部结构是引导程序运行及应用的向导,也是程序员在进行程序编写与设计时需要掌握与熟悉的内容。因此,对程序的内部结构的介绍是必不可少的。这里以创建名为 AndroidHello 的项目为例进行介绍。大家可以发现一个 App 程序是由多个文件及文件夹共同组成的,每个文件或文件夹都有不同的意义和功能。

在新建的 App 工程文件中,Android Studio 会自动生成许多文件,如图 3-1 所示。其中,重要的文件如下。

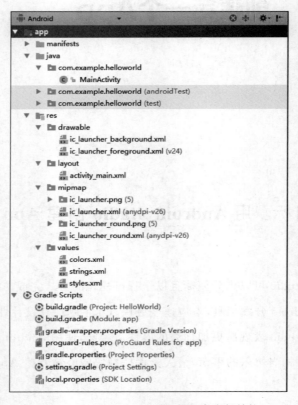

图 3-1　AndroidHello 项目的程序内部结构

（1）app：在 Android Studio 中进行编程时,一般分为 Project（工作空间）、Module（模块）两种概念。app 为创建项目时默认的模块,即一个 Module、一个 Android 应用程序的文档结构。

（2）libs：用于存放项目的类库,例如项目中会用到的.jar 文件等。

（3）src：用于存放该 Android 项目中用到的所有资源文件,例如图片等。

（4）androidTest：用于存放应用程序单元的测试代码。

（5）main：Android 项目的主目录,其中 java 目录存放.java 源代码文件,res 存放资源文件,包含图像、字符串资源等,AndroidManifest.xml 是项目的配置文件。

(6) build.gradle：Android 项目的 Gradle 构建脚本。

(7) build：Android studio 项目的编译目录。

(8) gradle：用于存放该项目的构建工具。

(9) External Libraries：用于显示该项目所依赖的所有类库。

3.2.2 Android 的开发流程

对 Android 平台上的应用进行开发，可以按照如下流程来进行。

(1) 安装 Android 调试软件，配置开发环境。

(2) 创建 Android 虚拟机或硬件设备。

(3) 创建 Android 项目，编写代码，提供资源文件。

(4) 运行 Android 应用程序，用 Android Studio 运行程序并呈现效果。

(5) 调试 Android 应用程序，测试并发布。

3.3 项目运行

3.3.1 创建一个 App

步骤 1：双击 Android Studio 的图标打开工具，如图 3-2 所示。

图 3-2　打开 Android Studio 时的加载画面

步骤2：单击 Start a new Android Studio project 按钮，创建第一个 Android 项目，如图 3-3 所示。

图 3-3　创建 Android 项目

步骤3：为新建项目设置属性。其中，项目的名称为 HelloWorld，设定项目位置，然后单击 Next 按钮，如图 3-4 所示。

步骤4：选择适当的 SDK 版本，然后单击 Next 按钮，如图 3-5 所示。

步骤5：为 App 的运行选择一个 Activity 样式，例如 Empty Activity，然后单击 Next 按钮，如图 3-6 所示。

步骤6：设置 Activity 的名称和 Layout 的名称，然后单击 Finish 按钮，完成新建程序的前期设定，如图 3-7 所示，继而进入 Android Studio 的开发主界面，如图 3-8 所示。

3.3.2　运行 App

1. 使用模拟器运行 App

为了避免系统报错，建议大家在创建模拟器之前单击 SDK Manager 按钮，更新 Android SDK 的相关配置，如图 3-9 所示。

步骤1：单击 AVD Manager 按钮，弹出 Android Virtual Device Manger 窗口，如图 3-10 所示，然后单击 `+ Create Virtual Device...` 按钮，创建虚拟设备模拟器。

项目3　创建第一个App

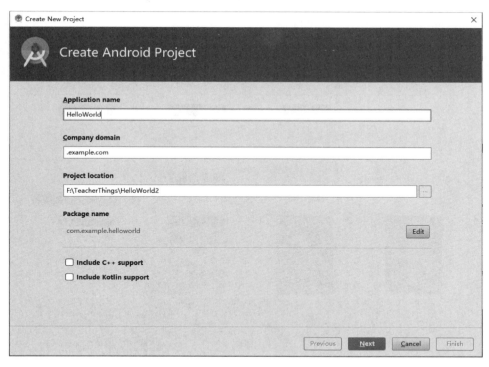

图 3-4　为新建项目设置属性

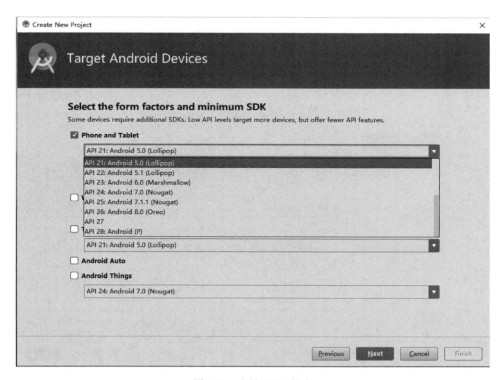

图 3-5　选择 SDK 版本

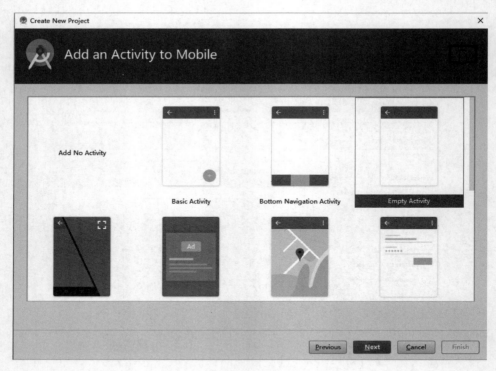

图 3-6 设置 App 的 Activity 样式

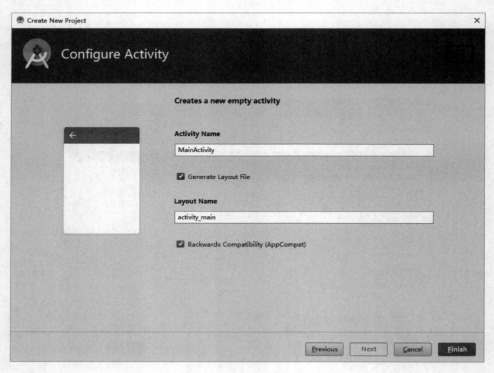

图 3-7 为新建 Activity 设置文件名称

项目3　创建第一个App

图 3-8　Android Studio 的开发主界面

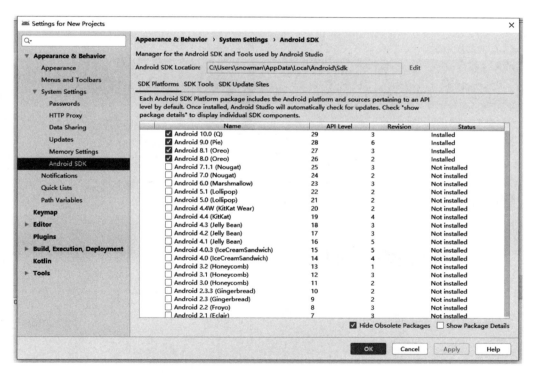

图 3-9　更新 Android Studio 的相关配置

步骤 2：选择一种模拟器设备，如图 3-11 所示。

步骤 3：确定设备后单击 Next 按钮，弹出 System Image 界面，选择系统版本，一般默

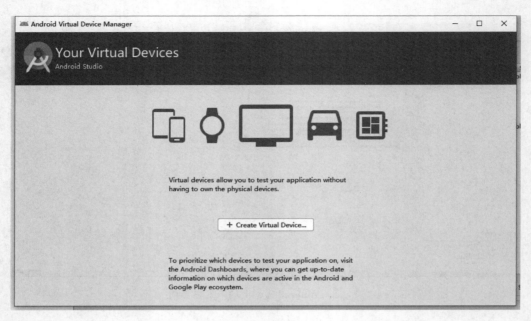

图 3-10　Android Virtual Device Manager 窗口

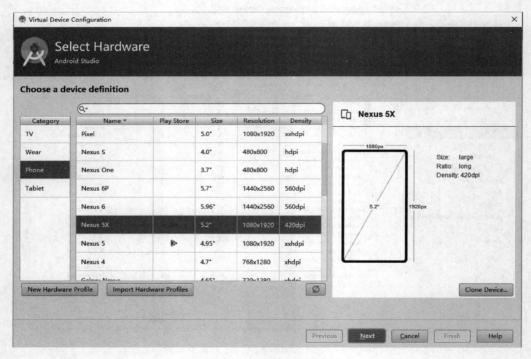

图 3-11　选择模拟器设备

认选择最高版本，如图 3-12 所示。

步骤 4：单击 Next 按钮，弹出 Verify Configuration 界面，一般应用默认设置。需要注

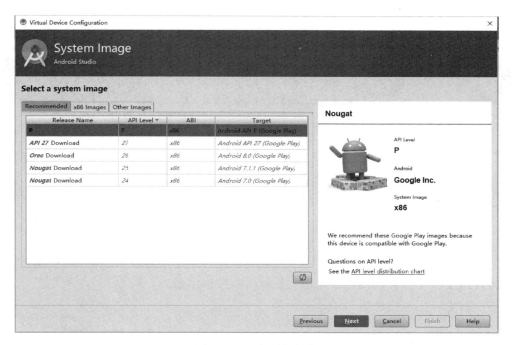

图 3-12　选择系统版本

意手机模拟器外观，一般将其设置为 Automatic Emulated Performance，即自动配置大小，如图 3-13 所示。然后单击 Finish 按钮。

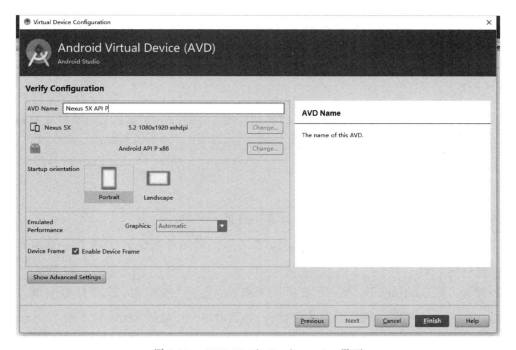

图 3-13　AVD Verify Configuration 界面

步骤 5：加载一段时间后，模拟器即被创建出来，如图 3-14 所示。

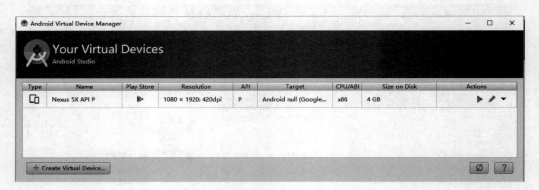

图 3-14　新创建的模拟器

步骤 6：单击 Run App 按钮，即可弹出选择部署目标 Select Deployment Target 窗口，再选择可用虚拟设备 Available Virtual Device，如图 3-15 所示。

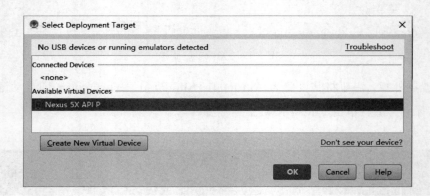

图 3-15　可用模拟器设备界面

步骤 7：单击 OK 按钮运行 App，可以通过 Android Studio 底部显示的情况来监视运行状态，加载一段时间后，即可出现使用模拟器运行的效果，如图 3-16 所示。

2. 使用手机运行 App

步骤 1：使用 USB 的方式将手机与计算机连接。

步骤 2：在手机上选中"连接后启动调试模式"，如果是首次连接，计算机会弹出安装对话框进行匹配安装。

步骤 3：在 Android Studio 的工具栏上单击"运行"按钮 ▶，选择部署目标 Select Deployment Target 窗口，选择要连接的设备，然后单击 OK 按钮，如图 3-17 所示。

图 3-16　使用模拟器运行的效果

图 3-17　连接设备

步骤 4：等待运行，即可在手机界面上看到 App 的演示效果，如图 3-18 所示。

3.3.3　调试 App

在使用 Android Studio 开发 App 的过程中，程序人员可能会由于各种原因在编程时产生错误或疏忽，因此通过调试的方法找到错误所在并进行修改是一项十分重要的工作。在使用 Android Studio 进行 App 开发时，可以采用断点调试的方法，具体操作过程如下。

步骤 1：在 Android Studio 中找到需要调试的程序，如图 3-19 所示。

图 3-18 使用手机运行 App 的效果

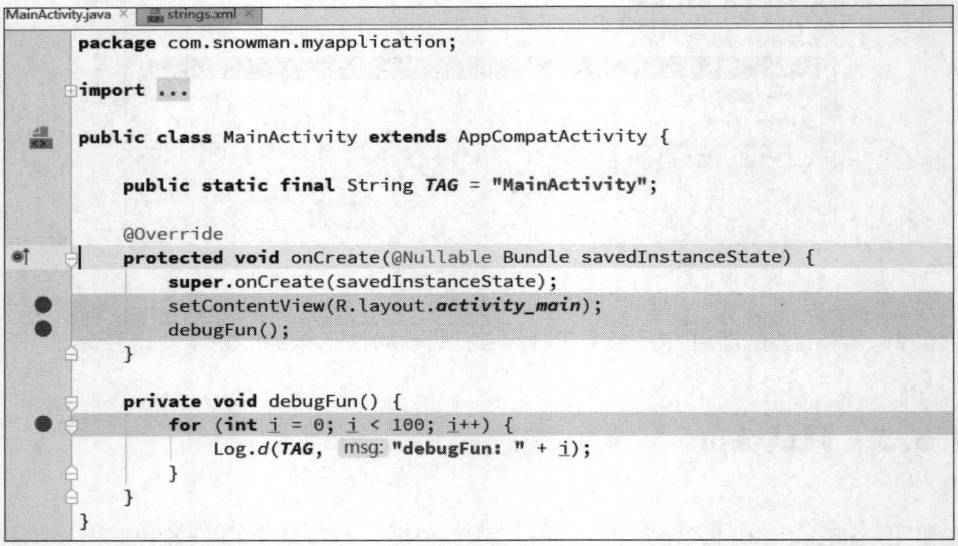

图 3-19 选择需要调试的程序

步骤 2：设置断点，即通过在行号处单击的方式进行设定。再单击■按钮，开启调试会话，在 Debug 视图中即可得到调试的情况，如图 3-20 所示。

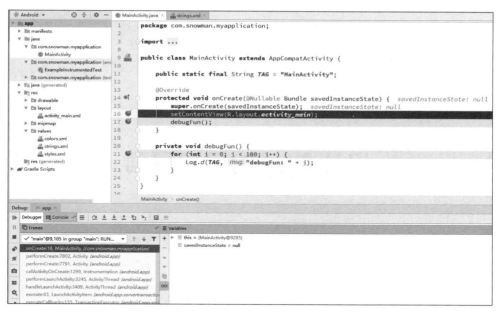

图 3-20　设置断点并开启调试会话

3.4　项目结案

本项目通过对 Android 内部结构的解析,让大家能够较为深入地体会到 App 的内部组成及彼此之间的关系;然后通过对 App 开发流程的梳理,明确开发 App 的全部步骤。从创建一个项目到运行项目,再到调试项目,整个流程紧密相关,缺一不可,为 App 的顺利编写提供了全面的技术保障。虽然在本项目中还没有接触到具体的编写方法,但是整体的流程与思路是大家需要掌握与熟练应用的。

3.5　项目练习

1. 创建一个名为"我的第一个 App"的小程序,体验完整的开发流程。
2. 在上一个小程序的基础上修改显示字符串为"Hello App"。
3. 在上一个小程序的基础上修改字符串的布局位置至屏幕左侧。

项目4

设计App的用户界面

4.1 项目目标:通过视图创建App的用户界面

App的交互界面是手机应用程序借助于手机终端与用户接触的直观显示与交互媒介,一款手机应用程序的设计成功与否,其交互界面的设计起着决定性的作用。一个好的设计,是让用户能够在使用应用程序的过程中体验到便捷,在交互的过程中体验到乐趣,而这些都是要通过编写代码实现的。本项目深入手机App交互界面的设计元素,从较为常用的交互界面设计方法与制作流程方面为大家详细介绍。

4.2 项目准备

在手机App的交互界面设计中,有3个常用的元素贯穿在整个设计与制作之中,分别是View(视图)、Widget(小部件)和ViewGroup(视图容器)。将三者合理地运用与设计,可以使手机交互界面更加人性化、多样化,如图4-1所示。

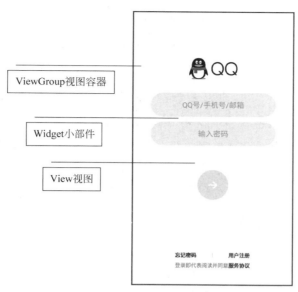

图 4-1　手机 App 界面各元素示意图

4.2.1　介绍视图类

1. View 元素

手机 App 是以视觉的方式向用户展现的,而可视范围即手机的屏幕区域。因此,所有的界面设计都要基于屏幕的大小、比例等因素。在手机界面上,通常可以看到各种按钮、文本框、布局效果、图片等可视化元素,这些都要通过 View 元素来实现。

View 在 Android SDK 中既是一个包,又是一个类,在应用的过程中多指其作为类时的功能。只有设置了 View 的各个属性,才能将 Widget 元素和 ViewGroup 元素放置在上面,显示效果。对于本项目将要介绍的各类组件来说,View 类是它们的父类,在应用与设计时需通过 View 类进行导入。

2. Widget 元素

Widget 元素是在 View 上需要实现的各个"小部件",例如 EditText(文本编辑)、ScrollView(滚动视图)、Button(按钮)等,它们的设计都是通过 Android SDK 中的 android.widget 包来实现的。

3. ViewGroup 元素

上述 View 元素和 Widget 元素中的视图内容需要依托于容器才能显示,而在 Android

程序中,Activity 就是可以盛放各类元素的容器,但其本身并不显示。因此,真正的界面元素要通过 Android SDK 中的 android.widget.ViewGroup 类来实现,它可以容纳各类部件,完成界面的布局。

ViewGroup 是 View 类的抽象子类,它的实现可以通过 Layout 布局来完成。在 ViewGroup 中有 FrameLayout 类用来控制屏幕区域中显示的单个项目;有 LinearLayout 类用来安排其子类出现在一列或单列位置上;有 RelativeLayout 类可以设置其子类的位置与其他子类或父类形成关联性布局等。

4.2.2 介绍资源文件夹

在 Android 的应用程序中,文字、图片、颜色、尺寸等资源需要与程序绑定,即将这些资源放置在配置文件中实现此种关联属性。下面将资源文件夹下的多种属性资源进行梳理,以供 App 界面设计时进行调用。

1. 字符串资源

在创建 App 时,在 Android Studio 窗口左侧的包中会呈现出整个程序的结构。字符串资源文件就位于 res 文件夹下的 values 文件夹中,即 strings.xml,如图 4-2 所示。双击打开

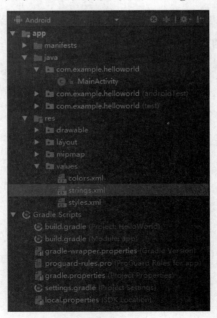

图 4-2　字符串资源文件的结构位置

strings.xml 文件,能够看到本程序中对于字符串的设定情况。其中,<resources></resources>标记为根元素,并且使用<string></string>标记对字符串进行定义;string name 后面填写字符串的名称,字符串的具体内容填写在<string>与</string>两个标记之间。

strings.xml 中的代码编写如下:

```
<?xml version = "1.0" encoding = "utf-8" standalone = "no"?>
<resources>
    <string name = "app_name">1</string>
    <string name = "hello world">Hello world!</string>
    <string name = "action_settings">Settings</string>
</resources>
```

字符串的名称 ←——————————————→ 字符串的具体内容

在设置字符串内容和对字符串进行布局时,多采用下面这种方法:

(1) 选中 res 文件夹下 values 中的 strings.xml 文件,使用字符串资源的语法格式为"[<package>.]R.string.字符串名"。

(2) 通过 src 文件夹下的 MainActivity.java 获取字符串,使用方法为"getResources().getString(R.string.字符串名)"。

(3) 在布局定义组件时(常用方法),通过字符串资源为 android:text 指定属性,具体应用为"android:text="@string/introduce"/>"。

2. 颜色资源

在进行 App 界面设计时,颜色(Color)是一项十分重要的参数。颜色的恰当运用能够使得 App 的视觉呈现更具吸引力,并且在辅助用户对于 App 功能的理解与应用方面锦上添花。对于颜色的设定与使用,是程序员开发 App 必不可少的技能。

对于 Android 开发而言,内部所采用的色彩模型为 RGB 模型,即由红、绿、蓝三原色组成,同时还有透明度(Alpha)数值的设定,通常对颜色的定义有 4 种形式,即♯RGB、♯ARGB、♯RRGGBB、♯AARRGGBB。其中,A、R、G、B 的取值均为 0~f;AA、RR、GG、BB 的取值均为 00~ff。

颜色资源文件 colors.xml 位于 res 文件夹下的 values 中,如图 4-3 所示。双击打开 colors.xml 文件,能够看到本程序中对于颜色的设定情况。其中,<resources></resources>标记为根元素,并且使用<color></color>标记对颜色进行定义;color name 后面填写颜色的名称,颜色的具体设定填写在<color>与</color>两个标记之间。

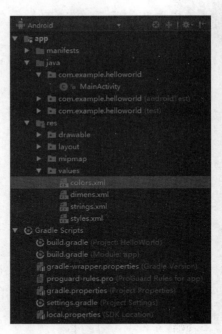

图 4-3 颜色资源文件所在的位置

colors.xml 中的代码编写如下：

```
< resources >
    < color name = "RGB">#f00</color>
    < color name = "ARGB">#70f0</color>
    < color name = "RRGGBB">#00ff00</color>
    < color name = "AARRGGBB">#6600ff00</color>
</resources>
```

颜色资源名称 ← → 具体颜色设定

在设置颜色资源名称和对颜色进行设定时，多采用下面这种方法：

(1) 在 Java 文件中使用颜色资源，基本语法为"[< package >.]R.color.颜色资源名称"。

(2) 在设置文字颜色时，通过 src 文件夹下的 MainActivity.java 获取颜色资源。以 TextView 组件为例，使用方法为：

```
TextView color = (TextView)findViewById(R.id.title);
color.setTextColor(getResources().getcolor(R.color.颜色资源名称));
```

(3) 在 XML 文件中使用颜色资源，基本语法为"@[< package >：]color/颜色资源名称"。

3. 尺寸资源

为了突出重点或者形成等级差异,对于文字、图形、图表等资源的尺寸进行设定是十分便捷的解决方法。在开发 App 的过程中,尺寸资源的设定与使用是其界面设计中的关键环节。

基于 Android 环境对 App 进行开发,常用的尺寸单位如表 4-1 所示。

表 4-1　Android 支持的尺寸单位及相关说明

单 位 表 示	单 位 名 称	单 位 说 明
px	像素	屏幕上的真实像素表示
in	英寸	基于屏幕的物理尺寸表示
mm	毫米	基于屏幕的物理尺寸表示
pt	点	基于字体尺寸表示
dp	和精度无关的像素	相对于屏幕物理密度的抽象单位
sp	和精度无关的像素	类似于 dp

(1) px:即像素(pixel),手机视图是由一个个像素点组成的。

(2) in:英寸,1in 约等于 2.54cm,主要用来描述手机屏幕的大小。

(3) pt:通常用来作为字体的尺寸单位,1pt 相当于 1/72in。

(4) dp(dip):即设备无关像素(device independent pixels),这种尺寸单位在不同设备上的物理大小相同。

(5) sp:它通常用作字体的尺寸单位,实际上其大小还与具体设备上的用户设定有关。

尺寸资源文件 dimens.xml 位于 res 文件夹下的 values 中,如图 4-4 所示。双击打开 dimens.xml 文件,能够看到本程序中对于尺寸的设定情况。其中,< resources ></resources >标记为根元素,并且使用< dimen ></dimen >标记对尺寸定义常量。dimen name 后面填写尺寸的名称,尺寸的具体数值设定填写在< dimen >与</dimen >两个标记之间。

dimens.xml 中的具体代码如下:

```
< resources >
    <!-- Default screen margins, per the Android Design guidelines. -->
    < dimen name = "activity_horizontal_margin">16dp</dimen>
    < dimen name = "activity_vertical_margin">16dp</dimen>
</resources>
```

尺寸资源名称　　　　　　　　　　　　具体尺寸大小

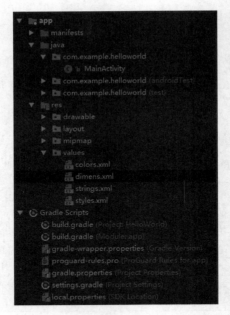

图 4-4　尺寸资源文件所在的位置

在设置尺寸资源名称与具体尺寸大小时,多采用下面这种方法:

(1) 在 Java 文件中使用尺寸资源,基本语法为"[< package >.]R.dimen.尺寸资源名称"。

(2) 在设置文字尺寸时,通过 src 文件夹下的 MainActivity.java 获取尺寸资源。以 TextView 组件为例,使用方法为:

```
TextView dimen = (TextView)findViewById(R.id.title);
dimen.setTextSize(getResources().getDimension(R.dimen.尺寸资源名称));
```

(3) 在 XML 文件中使用尺寸资源,基本语法为"@[< package >:]dimen/尺寸资源名称"。

4. 图片资源

在 App 的开发过程中,图片资源的使用是必不可少的。图像具有的形象性、生动性等特点是文字所无法比拟的。因此,制作一个优秀的 App 需要提供与其功能和使用方法相匹配的图形图像。

图片资源的来源比较广泛,在一般情况下,可以通过网络进行下载,或者自己绘制出各种格式的图形图像,例如.jpeg、.png、.gif 等常用的格式。当然,对于 Android 开发而言,在安装 SDK 时便自带制图工具 Draw 9-patch,如图 4-5 所示。该制图工具最大的优点在于所生成的图像具有相关性，即能够与手机屏幕的大小、方位等情况进行实时调整，且不失真，显示效果较好。该工具的使用方法如下：

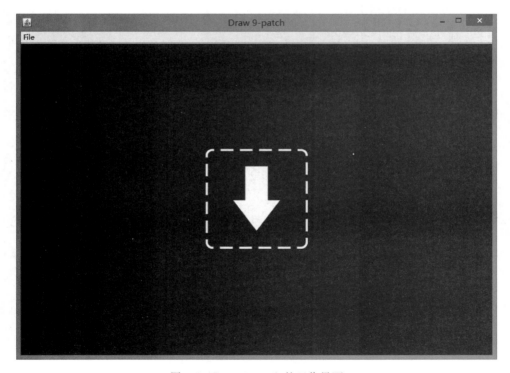

图 4-5　Draw 9-patch 工具的位置

步骤 1：Draw 9-patch 工具位于安装 SDK 的目录下的 tools 文件夹中。双击 draw9patch 图标，进入该工具的工作界面，如图 4-6 所示。

图 4-6　Draw 9-patch 的工作界面

步骤 2：导入图片。选择工作界面左上角的 File 命令，弹出一个菜单栏，其中有 Open 9-patch（打开图片）选项、Save 9-patch（保存图片）选项和 Quit（退出）选项，如图 4-7 所示。

选择 Open 9-patch 命令,选择本地计算机中已经保存好的图片,导入到 Draw 9-patch 工具中。工作界面的左侧为导入图片的初始状态,可通过 Ctrl 与 Shift 键配合鼠标左键对其选用范围进行调整;工作界面的右侧为该图像的拉伸预览图,分别为纵向拉伸、横向拉伸和整体拉伸,如图 4-8 所示。

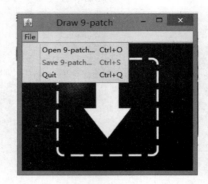

图 4-7 Draw 9-patch 的菜单栏

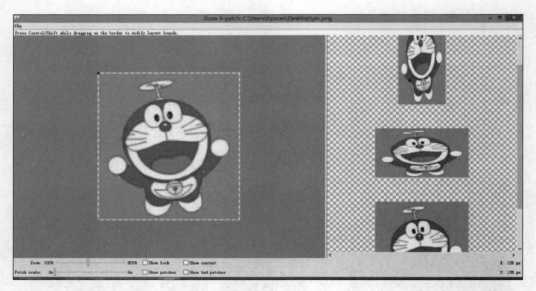

图 4-8 导入图片

步骤 3:保存图片。选择 File|Save 9-patch 命令,将图片另存到指定地址,并重新命名,即可生成扩展名为 .9.png 的图片,它可成为 App 开发时的图像资源,并通过程序调用。

注意:扩展名为 .9.png 和 9-patch 的图片均可成为 App 开发时的图像资源。

在引用图像资源与对图像进行设定时,多采用下面这种方法。

(1) 在 Java 文件中访问图像资源,基本语法为"[< package >.]R.drawable.<文件名称>"。

(2) 在设置图像显示时,通过 src 文件夹下的 MainActivity.java 获取图像资源。以

ImageView 组件为例,使用方法为:

```
ImageView pic = (ImageView)findViewById(R.id.ImageView1);
pic.setImageResource(R.drawable.head);
```

(3) 在 XML 文件中使用图像资源,基本语法为"@[<package>:]drawable/文件名称"。

4.2.3 介绍布局类

为了 App 的 UI 组件设计适应于不同型号手机的屏幕大小与比例,Android 提供了不同类型的布局方式。在创建 App 时,在 Android Studio 窗口左侧的项目模块中会呈现出整个程序的结构。布局文件就位于 res 文件夹下的 layout 文件夹中,即 main.xml,如图 4-9 所示。

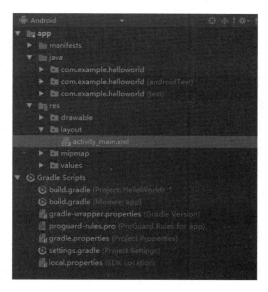

图 4-9 布局资源文件的结构位置

双击打开 main.xml 文件,能够看到本程序中对于布局的设定情况。对于布局资源的设定,包括 AbsoluteLayout(绝对布局)、LinearLayout(线性布局)、RelativeLayout(相对布局)、TableLayout(表格布局)、FrameLayout(帧布局)、GridLayout(网格布局)6 种布局方式。其中,AbsoluteLayout 是通过设定 x、y 的数值为文字、图片等资源进行位置设定,虽然十分精确,但考虑到手机屏幕尺寸的多样性与屏幕翻转的应用性,这种布局方式逐渐显示出其局限性;LinearLayout 较 AbsoluteLayout 适用性更强,这种方式会使系统以顺序动态的方式将文字、图片等资源放置在屏幕上,但是 LinearLayout 只能设定水平和垂直中的一种方向,因此仍然具有较大的局限性;RelativeLayout 则是适用性最高、最为人性化的布局方式,它能够通过定义相对位置很好地将子视图动态地固定在其父视图的相关位置,且对于手机屏幕的通用性与屏幕翻转的变化性有优秀的表现;GridLayout 是 Android 4.0(API Level 14)新引入的网格矩阵形式的布局控件。

在访问布局资源时,多采用下面两种方法。

(1) 在 Java 开发中,使用下面布局资源的语法格式进行访问:

[<package>.]R.layout.<文件名称>

(2) 在 XML 文件中,使用下面布局资源的语法格式进行访问:

@[<package>:]layout.文件名称

1. AbsoluteLayout(绝对布局)

<AbsoluteLayout></AbsoluteLayout>标记为根元素,具体的语法格式为:

```
<AbsoluteLayout xmlns:android="http://schemas.android.com/apk/res/android"
属性列表
>
</AbsoluteLayout>
```

其常用的基本属性如下。

android:layout_x:用于设定当前子类控件的 x 位置。

android:layout_y:用于设定当前子类控件的 y 位置。

2. LinearLayout(线性布局)

<LinearLayout></LinearLayout>标记为根元素,具体的语法格式为:

```
<LinearLayout xmlns:android="http://schemas.android.com/apk/res/android"
属性列表
>
</LinearLayout>
```

其常用的基本属性如下。

(1) android:orientation:用于设定水平排列或垂直排列。vertical 为垂直排列(也为默认值),horizontal 为水平排列。

(2) android:gravity:用于设定放置的方向。center 为水平、垂直皆居中;center_vertical 为垂直居中;center_horizontal 为水平居中;top 为置顶;bottom 为置底;left 为置

左；right 为置右。

（3）android：layout_weight：用于设定子元件或子框架的比重。

（4）android：layout_width：用于设定组件的宽度。match_parent 为组件的宽度与父容器匹配；wrap_content 为组件的宽度刚好能够包容其中内容，例如文字或图片等。

（5）android：layout_height：用于设定组件的高度。

（6）android：id：用于为组件设定 id 属性。

（7）android：background：用于设定背景。

3. RelativeLayout（相对布局）

< RelativeLayout ></RelativeLayout >标记为根元素，具体的语法格式为：

```
< RelativeLayout xmlns : android = "http://schemas.android.com/apk/res/android"
属性列表
>
</RelativeLayout >
```

其常用的基本属性如下。

（1）android：gravity：用于设定布局中各个组件间的对齐方式。

（2）android：ignoreGravity：用于设定布局中哪个组件不受 gravity 属性制约。

4. TableLayout（表格布局）

< TableLayout ></TableLayout >和< TableRow ></TableRow >标记为根元素，具体的语法格式为：

```
< TableLayout xmlns: android = "http://schemas.android.com/apk/res/android"
属性列表
>
< TableRow 属性列表>UI 组件</TableRow>
</RelativeLayout >
```

其常用的基本属性如下。

（1）android：collapseColumns：用于设定隐藏的列。

（2）android：stretchColumns：用于设定可伸展的列。该列可以向行方向伸展，最多可占据一整行。

（3）android：shrinkColumns：用于设定可收缩的列。当该列子控件的内容太多，已经挤满所在行时，该子控件的内容将往列方向显示。

5. FrameLayout(帧布局)

<FrameLayout></FrameLayout>标记为根元素,具体的语法格式为:

```
< FrameLayout xmlns: android = "http://schemas.android.com/apk/res/android"
属性列表
>
</FrameLayout>
```

其常用的基本属性如下。

(1) android:foreground:用于设定前景图像。

(2) android:foregroundGravity:用于设定前景图像的位置信息。

6. GridLayout(网格布局)

<GridLayout></GridLayout>标记为根元素,具体的语法格式为:

```
< GridLayout xmlns: android = "http://schemas.android.com/apk/res/android"
属性列表
>
</GridLayout>
```

其常用的基本属性如下。

(1) android:columnCount:用于设定最多的列数,为整数。

(2) android:rowCount:用于设定最多的行数,为整数。

4.3 项目运行

4.3.1 字符串资源

【例 4.1】 创建一个手机 App 项目,将其命名为 4.1,通过对字符串的设置与设计,制作一个"美丽新世界"App 的简介界面。

步骤 1:在 Android Studio 中导入名为"4.1"的项目文件夹,选择 res 文件夹下 values 中的 strings.xml 文件,可见其中对于字符串的命名分别为 title、introduce、company、url,具体代码如下:

视频讲解

```xml
<resources>
    <string name="app_name">4.1 字符串资源案例</string>
    <string name="title">美丽新世界</string>
    <string name="introduce">《美丽新世界》是英国作家阿道司·赫胥黎创作的长篇小说.《美丽新世界》的故事背景设在福特纪元632年(即遥远的未来2532年)的人类社会.它是世界性国家,被称之为"文明社会",之外还有"蛮族保留区",由一些印第安部落居住.伯纳和列宁娜去保留区游览时遇到了约翰和他的母亲琳达,琳达曾是新世界的居民,来游玩时不慎坠下山崖而留下,并生下了约翰.伯纳出于自己的目的,将琳达母子带回新世界,琳达很快因服用过量的唆麻(一种兴奋剂)而死,约翰对新世界也由崇拜转为厌恶,与新世界激烈冲突后自缢身亡.</string>
    <string name="company">作品出处: Chatto and Windus</string>
    <string name="url">作品网址: https://baike.sogou.com</string>
</resources>
```

步骤2:选择res文件夹下layout中的main.xml文件,可见其中对于上述4个字符串资源进行了字体、位置、布局等属性的设定,具体代码如下:

```xml
<TextView
    android:text="@string/title"
    android:gravity="center"
    android:layout_width="match_parent"
    android:layout_height="wrap_content"
/>

<TextView
    android:layout_width="wrap_content"
    android:layout_height="wrap_content"
    android:text="@string/introduce"
/>

<TextView
    android:layout_width="match_parent"
    android:layout_height="wrap_content"
    android:gravity="center"
    android:text="@string/company"
/>

<TextView
    android:text="@string/url"
    android:gravity="center"
    android:layout_width="match_parent"
    android:layout_height="wrap_content"
/>
```

步骤3：运行该项目，手机模拟器显示效果如图 4-10 所示。

4.3.2 颜色资源

基于例 4.1，分别为该 App 界面上所呈现的 4 类文字进行颜色的设定。

步骤1：在 Android Studio 中导入名为"4.1"的项目文件夹，选择 res 下的 values 文件夹，新建名为 colors.xml 的文件，并为 title、introduce、company、url 的颜色进行设定，具体代码如下：

图 4-10 字符串资源设置的运行效果

```xml
<?xml version = "1.0" encoding = "utf-8"?>
<resources>
    <color name = "title">#ff0</color>
    <color name = "introduce">#0f0</color>
    <color name = "company">#0ff</color>
    <color name = "url">#908f</color>
</resources>
```

步骤2：选择 res 文件夹下 layout 中的 main.xml 文件，添加对于上述 4 个颜色资源的设定，具体代码如下：

```xml
<TextView
    android:text = "@string/title"
    android:textColor = "@color/title"
    android:gravity = "center"
    android:layout_width = "match_parent"
    android:layout_height = "wrap_content"
/>

<TextView
    android:layout_width = "wrap_content"
    android:layout_height = "wrap_content"
    android:text = "@string/introduce"
    android:textColor = "@color/introduce"
/>

<TextView
    android:layout_width = "match_parent"
    android:layout_height = "wrap_content"
```

```
        android:gravity = "center"
        android:text = "@string/company"
        android:textColor = "@color/title"
    />

    <TextView
        android:text = "@string/url"
        android:gravity = "center"
        android:textColor = "@color/url"
        android:layout_width = "match_parent"
        android:layout_height = "wrap_content"
    />
```

步骤3：运行该项目，手机模拟器显示效果如图4-11所示。

图4-11 颜色资源设置的运行效果

4.3.3 尺寸资源

基于例4.1，分别为该App界面上所呈现的4类文字进行尺寸大小的设定。

步骤1：在Eclipse中导入名为"4.1"的项目文件夹，选择res下的values文件夹，新建名为dimens.xml的文件，并为title、introduce、padding、titlePadding的尺寸进行设定，具体代码如下：

```
<?xml version = "1.0" encoding = "utf-8"?>
<resources>
    <dimen name = "title">30dp</dimen>
    <dimen name = "introduce">16dp</dimen>
    <dimen name = "padding">10dp</dimen>
    <dimen name = "titlePadding">10dp</dimen>
</resources>
```

步骤2：选择res文件夹下layout中的main.xml文件，添加对于上述4个尺寸资源的设定，具体代码如下：

```
    <TextView
        android:text = "@string/title"
        android:padding = "@dimen/titlePadding"
        android:textSize = "@dimen/title"
        android:textColor = "@color/title"
        android:gravity = "center"
        android:layout_width = "match_parent"
        android:layout_height = "wrap_content"
```

```
    />
<TextView
    android:layout_width = "wrap_content"
    android:layout_height = "wrap_content"
    android:text = "@string/introduce"
    android:textColor = "@color/introduce"
    android:textSize = "@dimen/introduce"

    />

<TextView
    android:layout_width = "match_parent"
    android:layout_height = "wrap_content"
    android:gravity = "center"
    android:padding = "@dimen/padding"
    android:text = "@string/company"
    android:textColor = "@color/title"
    />
<TextView
    android:text = "@string/url"
    android:gravity = "center"
    android:textColor = "@color/url"
    android:paddingLeft = "@dimen/padding"
    android:layout_width = "match_parent"
    android:layout_height = "wrap_content"
    />
```

图 4-12　尺寸资源设置的运行效果

步骤 3：运行该项目，手机模拟器显示效果如图 4-12 所示。

4.3.4　图片资源

【例 4.2】　创建一个手机 App 项目，将其命名为 4.2，通过对图像的调用与设置，制作一个带有按钮的 App 界面，对图像的调用方式分别为调用普通图片、调用 9-patch 图像并设置精确尺寸、调用 9-patch 图像并设置交互效果。

步骤 1：准备 3 张按钮图片，分别为 red.png、green.png 和 yellow.png，将 green.png 和 yellow.png 两张图片用 Draw 9-patch 进行另存为操作，使其扩展名变更为 green.9.png 和 yellow.9.png，如图 4-13 所示。

视频讲解

步骤 2：在 Eclipse 中导入名为 "4.2" 的项目文件夹，选择 res 文件夹下 values 中的 strings.xml 文件，对 3 个按钮上的文字进行设定，具体代码如下：

图 4-13　3 张备用图像资源

```xml
<resources>
    <string name = "hello">Hello World, MainActivity!</string>
    <string name = "app_name">4.2 图片资源</string>
    <string name = "pic1">普通图片 PNG</string>
    <string name = "pic2">9-patch(设置尺寸)</string>
    <string name = "pic3">9-patch(有交互效果)</string>
</resources>
```

步骤3：选择 res 文件夹下 layout 中的 main.xml 文件，可见其中对于3个按钮的定义及图片的导入，具体代码如下：

```xml
<Button
    android:id = "@+id/button1"
    android:background = "@drawable/red"
    android:layout_margin = "5dp"
    android:layout_width = "match_parent"
    android:layout_height = "50dp"
    android:text = "@string/pic1"
    />

<Button
    android:id = "@+id/button2"
    android:layout_width = "match_parent"
    android:layout_height = "20dp"
    android:layout_margin = "5dp"
    android:background = "@drawable/green"
    android:text = "@string/pic2"
    />

<Button
    android:id = "@+id/button3"
    android:background = "@drawable/button_state"
    android:layout_margin = "5dp"
    android:layout_width = "match_parent"
    android:layout_height = "wrap_content"
    android:text = "@string/pic3"
    />
```

步骤4：设定具有交互效果的按钮功能，在 drawable 目录的上级目录(res 或更上级的目录)右击，选择 New|Android resource file 选项，如图4-14所示。

步骤5：在弹出的 New Resource File 对话框中选择 Drawable 资源类型，并输入要创建的文件类型，如图4-15所示。

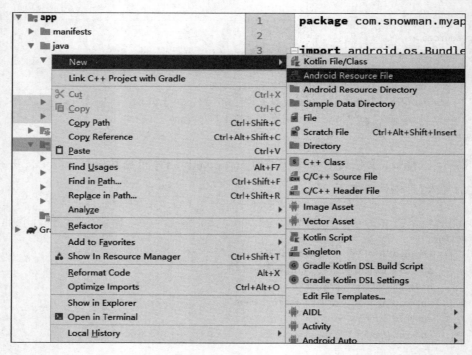

图 4-14 选择 Android resource file 选项

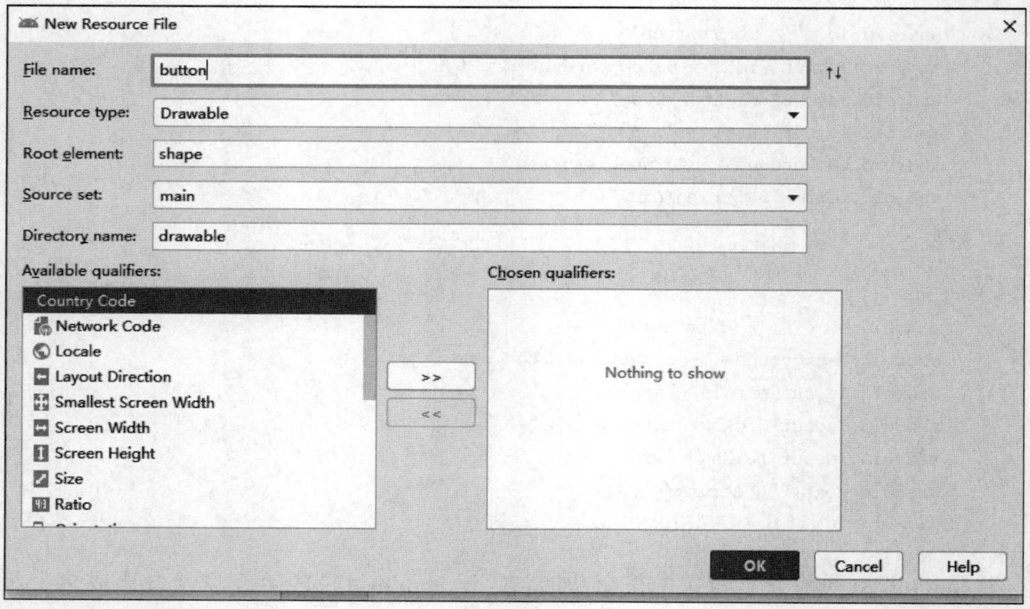

图 4-15 设置按钮

步骤6：运行程序，得到如图4-16所示的效果。其中，普通图片PNG按钮为直接引用本地计算机的PNG图片，因与手机屏幕大小不匹配，所以有失真的现象出现；9-patch（设置尺寸)按钮和9-patch(有交互效果)按钮因导成.9.patch格式的图片,所以没有失真的现象,并且9-patch(设置尺寸)按钮因重新设定了宽高像素值而变形,9-patch(有交互效果)按钮能够在单击时变为普通图片PNG按钮。

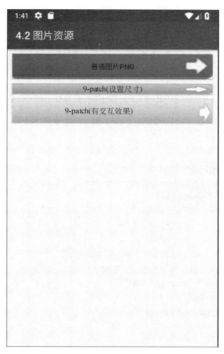

图 4-16　不同类型图片资源的演示效果

4.3.5　布局类

【例 4.3】 创建一个手机 App 项目,将其命名为 4.3,应用 UI 设计中关于布局、资源文件的应用等方法实现《植物大战僵尸》游戏的欢迎界面。

步骤 1:准备图像素材。其中包括游戏欢迎界面的背景图片 bg1.png 和 bg2.png,以及"开始"按钮图片 play.png、"停止"按钮图片 stop.png、"帮助"按钮图片 help.png、"分享"按钮图片 share.png 等,如图 4-17 所示。将上述素材放置到 res 下的 drawable-mdpi 文件夹中。

视频讲解

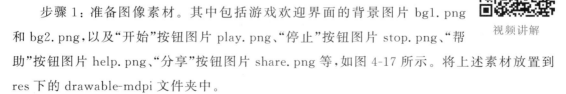

图 4-17　App 界面视图素材

步骤2：定义和设计手机界面的基本文字状态。新建名为4.3的Android开发包，在res下values文件夹中的strings.xml文件中修改初始代码，具体代码如下：

```xml
<?xml version = "1.0" encoding = "utf-8" standalone = "no"?>
<resources>
    <string name = "hello">植物大战僵尸</string>
    <string name = "app_name">4.3 UI 界面设计</string>
</resources>
```

步骤3：用线性布局、相对布局和网格布局的方式定义素材及相关属性。在res下layout文件夹中的main.xml文件中修改代码，具体代码如下：

```xml
<?xml version = "1.0" encoding = "utf-8"?>
<LinearLayout xmlns:android = "http://schemas.android.com/apk/res/android"
    android:orientation = "vertical"
    android:layout_width = "fill_parent"
    android:layout_height = "fill_parent">

    <ImageView android:layout_width = "match_parent"
        android:layout_height = "wrap_content"
        android:scaleType = "centerCrop"
        android:layout_weight = "1"
        android:src = "@drawable/bg1"
        android:contentDescription = "@null"/>

    <RelativeLayout
        android:id = "@+id/relativeLayout1"
        android:layout_width = "match_parent"
        android:layout_height = "wrap_content"
        android:layout_weight = "2"
        android:background = "@drawable/bg2"
        android:contentDescription = "@null" >

        <GridLayout
            android:id = "@+id/gridLayout1"
            android:layout_width = "wrap_content"
            android:layout_height = "wrap_content"
            android:layout_above = "@+id/imageButton2"
            android:layout_alignParentLeft = "true"
            android:layout_marginLeft = "42dp" >
        </GridLayout>

        <ImageView
            android:id = "@+id/ImageView01"
            android:layout_width = "wrap_content"
            android:layout_height = "wrap_content"
            android:layout_alignParentLeft = "true"
            android:layout_below = "@+id/imageButton0"
            android:layout_marginTop = "33dp"
```

```xml
            android:contentDescription = "@null"
            android:src = "@drawable/help" />
        < ImageView
            android:id = "@ + id/ImageView02"
            android:layout_width = "wrap_content"
            android:layout_height = "wrap_content"
            android:layout_alignParentRight = "true"
            android:layout_alignTop = "@ + id/ImageView01"
            android:contentDescription = "@null"
            android:src = "@drawable/share" />

        < ImageView
            android:id = "@ + id/imageButton0"
            android:layout_width = "wrap_content"
            android:layout_height = "wrap_content"
            android:layout_alignParentTop = "true"
            android:layout_marginTop = "23dp"
            android:layout_toRightOf = "@ + id/ImageView01"
            android:contentDescription = "@null"
            android:src = "@drawable/play" />

        < ImageView
            android:id = "@ + id/imageButton2"
            android:layout_width = "wrap_content"
            android:layout_height = "wrap_content"
            android:layout_alignTop = "@ + id/imageButton0"
            android:layout_marginLeft = "32dp"
            android:layout_toRightOf = "@ + id/imageButton0"
            android:contentDescription = "@null"
            android:src = "@drawable/stop" />

    </RelativeLayout >

</LinearLayout >
```

步骤4：用线性布局的方式定义素材及相关属性。在 src 下 com.mingrisoft 文件夹中的 MainActivity.java 文件中修改代码，具体代码如下：

```java
package com.mingrisoft;

import android.app.Activity;
import android.os.Bundle;
import android.view.View;
import android.view.View.OnClickListener;
import android.widget.ImageView;
import android.widget.Toast;

public class MainActivity extends Activity {
    @Override
    public void onCreate(Bundle savedInstanceState) {
        super.onCreate(savedInstanceState);
        setContentView(R.layout.main);
```

```
        addClick();
    }
    public void addClick(){
        ImageView img0 = (ImageView)findViewById(R.id.imageButton0);
        img0.setOnClickListener(new OnClickListener() {

            @Override
            public void onClick(View v) {
                // TODO Auto-generated method stub
                Toast.makeText(MainActivity.this,"开始游戏", Toast.LENGTH_SHORT).show();
            }
        });

         ImageView img2 = (ImageView)findViewById(R.id.imageButton2);
        img2.setOnClickListener(new OnClickListener() {

            @Override
            public void onClick(View v) {
                // TODO Auto-generated method stub
                Toast.makeText(MainActivity.this,"结束游戏", Toast.LENGTH_SHORT).show();
            }
        });

    }
}
```

步骤5：通过以上4步的设置，运行程序，得到 App 欢迎界面，如图4-18所示。

图 4-18　App 欢迎界面的视图效果

4.4 项目结案

通过本项目的图文讲解,向大家展示了如何对手机 App 进行界面设计与开发,其中包括 6 种布局管理器及其相关使用方法和属性、对于各种类型资源的定义与设置,以及相关的案例说明。希望大家在这样翔实的讲解下,能够掌握 App 界面开发的技术,并在此基础上多做练习,以达到举一反三的学习效果。

4.5 项目练习

1. 开发一款音乐 App 的开始界面,使用线性布局对相关素材进行设计。
2. 开发一款健康运动 App 的功能菜单界面,使用相对布局对按钮素材进行设计,并添加响应机制。
3. 开发一款阅读 App 的书架界面,使用表格布局对图像素材进行设计。

项目5

理解App的活动

5.1 项目目标:理解App的活动机制与状态

本项目的内容是创建和设计 App 必备的基础理论知识。任何一个 App 都像人的生命一样,有开始、运行、暂停和消亡,在这个过程中,可以用各种方法对程序进行控制。本项目设定的目标为掌握 Activity 的状态、生命周期及相关属性,熟练编写代码实现创建 Activity,设置 Activity 以及启动、关闭 Activity 等操作。

5.2 项目准备

Android 具有四大组件,分别为 Activity、Service(服务)、Content Provider(内容提供)、BroadcastReceiver(广播接收器)。Activity 是 Android 组件中最基本也是最常用的四大组件之一,其本质是一个应用程序组件,专门为人机交互提供可视化屏幕、交互接口,并完成交互任务。Activity 可以通过 setContentView(View)来显示指定控件。在一个 Android 应用中,一个 Activity 通常就是一个单独的屏幕,它上面可以显示一些控件,也可以监听并对用户的事件做出响应。Activity 之间通过 Intent 进行通信。

5.2.1 介绍 Activity 的状态

在 Android 中,Activity 拥有以下 4 种基本状态。

(1) Active/Running(运行状态):一个新 Activity 启动入栈后,它在屏幕最前端,处于栈的最顶端,此时它处于可见并可与用户交互的激活状态。

(2) Paused(暂停状态):当 Activity 被另一个透明或者 Dialog 样式的 Activity 覆盖时,处于 Paused 状态。此时它依然与窗口管理器保持连接,系统继续维护其内部状态,所以它仍然可见,但它已经失去了焦点,故不可与用户交互。

(3) Stopped(停止状态):当 Activity 被另外一个 Activity 覆盖、失去焦点并不可见时,处于 Stopped 状态。

(4) Killed(死亡状态):Activity 被系统杀死回收或者没有被启动时,处于 Killed 状态。

当一个 Activity 实例被创建、销毁或者启动另外一个 Activity 时,它在这 4 种状态之间进行转换,这种转换的发生依赖于用户程序的动作。图 5-1 说明了 Activity 在不同状态之间转换的时机和条件。

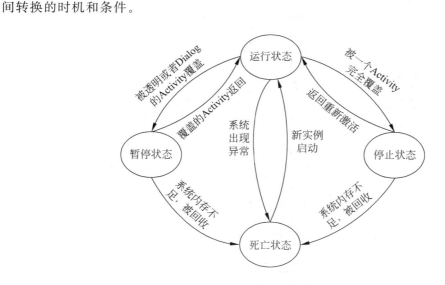

图 5-1 Activity 的 4 种状态的转换

值得注意的是,开发 Android 的程序员开启一个 Activity,却不能够通过 Activity.finish()方法结束它,仅能使其被回收,即从 Active/Running 状态转到 Paused 状态。

5.2.2 介绍 Activity 的生命周期

根据官方文档对 Activity 生命周期的说明(如图 5-2 所示),可以对任一款 App 的生命周期活动进行描述。首先,在第一次打开某款 App 时,当前的 FirstActivity 即被执行,随后 Android 系统依次调用 onCreate()方法、onStart()方法、onResume()方法,使得 FirstActivity 完全被启动,并显示到前台。如果 Intent 同时启动了 SecondActivity,则会通过 onPause()方法暂停之前的 FirstActivity。如果又选择了 Back 键,则会通过 onResume()方法重新返回到 FirstActivity,而 SecondActivity 则被 onStop()方法所销毁。如果想继续 SecondActivity,则需要通过 onRestart()方法来实现。如果内存不够,或者结束了 Activity,那么就会通过 onDestroy()方法结束整个 Activity,必须重新创建才能重新启动该 App。

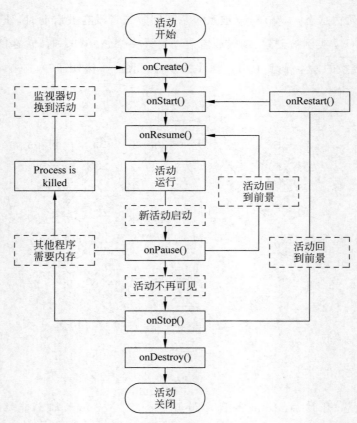

图 5-2 Activity 生命周期循环图

对于 Activity 生命周期而言,有 7 个方法能够帮助 Activity 进入循环流程。这 7 个方法分别为 onCreate()、onRestart()、onStart()、onResume()、onPause()、onStop() 和 onDestroy()。其具体的使用情境、详细说明以及方法之间的承接方法如表 5-1 所示。

表 5-1　Activity 生命周期方法及使用说明

生命周期所用方法	使 用 情 境	详 细 说 明	承 接 方 法
onCreate()	首次创建 Activity 时调用	用户应该在此方法中执行所有正常的静态设置,例如创建视图、将数据绑定到列表等。系统向此方法传递一个 Bundle 对象,其中包含 Activity 的上一状态,不过前提是捕获了该状态	始终后接 onStart()
onRestart()	在 Activity 已停止并即将再次启动前调用	Activity 被重新激活时,就会调用 onRestart 方法	始终后接 onStart()
onStart()	在 Activity 即将对用户可见之前调用	onStart() 是 Activity 界面被显示出来的时候执行的	如果 Activity 转入前台,则后接 onResume();如果 Activity 转入隐藏状态,则后接 onStop()
onResume()	在 Activity 即将开始与用户进行交互之前调用	此时,Activity 处于 Activity 堆栈的顶层,并具有用户输入焦点	始终后接 onPause()
onPause()	当系统即将开始继续另一个 Activity 时调用	此方法通常用于确认对持久性数据的未保存更改、停止动画以及其他可能消耗 CPU 的内容,诸如此类。它应该非常迅速地执行所需操作,因为它返回后,下一个 Activity 才能继续执行	如果 Activity 返回前台,则后接 onResume();如果 Activity 转入对用户不可见状态,则后接 onStop();如果它在后台仍然可见,则不会停止
onStop()	在 Activity 对用户不再可见时调用	如果 Activity 被销毁,或另一个 Activity(一个现有 Activity 或新 Activity)继续执行并将其覆盖,就可能发生这种情况	如果 Activity 恢复与用户的交互,则后接 onRestart();如果 Activity 被销毁,则后接 onDestroy()
onDestroy()	在 Activity 被销毁前调用	这是 Activity 将收到的最后调用。当 Activity 结束(有人对 Activity 调用了 finish()),或系统为节省空间而暂时销毁该 Activity 实例时,可能会调用它。用户可以通过 isFinishing() 方法区分这两种情形	此方法为生命周期的结束方法

根据 Activity 官方文档说明，Activity 的生命周期有 3 个重要的循环对。

（1）整个生命周期：从调用 onCreate()方法直到调用 onDestroy()方法的整个过程。Activity 需要通过 onCreate()方法准备启动全局状态，再通过 onDestroy()方法释放所有的资源。例如，当一个线程在后台通过网络下载数据时，则通过 onCreate()方法创建线程，再通过 onDestroy()方法停止线程。

（2）可见生命周期：从调用 onStart()方法到调用 onStop()方法之间的过程。之所以称之为可见生命周期，是因为用户可以直观地在屏幕上看到这个过程，并且能够保持需要展示给用户的资源。例如，通过调用 onStart()方法注册一个 BroadcastReceiver，用于监视使 UI 产生变化的内容，再通过调用 onStop()方法取消监视。onStart()方法和 onStop()方法可以在被用户可见和隐藏两种方式切换的时候被多次调用。

（3）前台生命周期：从调用 onResume()方法到调用 onPause()方法之间的过程。在此过程中，Activity 通过前台与用户产生交互。Activity 从 onResume()到 onPause()进行了十分频繁的切换。例如，当设备处于休眠状态时，将调用 onPause()方法；当 result 和一个新的 Intent 发送给 Activity 时，将调用 onResume()方法。

5.2.3　介绍 Activity 的属性

Activity 作为 Android 的对象，需要通过各种属性的设置来实现。Activity 的属性类型多样，表 5-2 列出了相关属性及其描述和调用方法。

表 5-2　Activity 的属性详表

属　　性	描　　述	调　用　方　法
android：allowTaskReparenting	是否允许 Activity 更换从属的任务，例如从短信息任务切换到浏览器任务	android：allowTaskReparenting = ["true"|"false"]
android：alwaysRetainTaskState	是否保留状态不变，例如切换回 home，再重新打开，Activity 处于最后的状态	android：alwaysRetainTaskState = ["true"|"false"]
android：clearTaskOnLaunch	例如 P 是 Activity，Q 是被 P 触发的 Activity，然后返回 Home，重新启动 P，是否显示 Q	android：clearTaskOnLaunch = ["true"|"false"]

续表

属　性	描　述	调用方法
android：configChanges	当配置 List 发生修改时，是否调用 onConfigurationChanged（）方法，例如" locale ︱ navigation ︱ orientation"	android：configChanges＝［oneormoreof："mcc""mnc""locale""touchscreen""keyboard""keyboardHidden""navigation""orientation""fontScale"］
android：enabled	Activity 是否可以被实例化	android：enabled＝["true"︱"false"]
android：excludeFromRecents	是否可被显示在最近打开的 Activity 列表里	android：excludeFromRecents＝["true"︱"false"]
android：exported	是否允许 Activity 被其他程序调用	android：exported＝["true"︱"false"]
android：finishOnTaskLaunch	是否关闭已打开的 Activity（当用户重新启动这个任务的时候）	android：finishOnTaskLaunch＝["true"︱"false"]
android：icon	调用图标	android：icon＝"drawableresource"
android：label	调用标签	android：label＝"stringresource"
android：launchMode	Activity 启动方式："standard""singleTop""singleTask""singleInstance" 其中前两个为一组，后两个为一组	android：launchMode＝["multiple"︱"singleTop"︱"singleTask"︱"singleInstance"]
android：multiprocess	可以多实例	android：multiprocess＝["true"︱"false"]
android：name	采用类名的简写方式，查看文档类名的简写格式	android：name＝"string"
android：noHistory	是否需要移除这个 Activity（当用户切换到其他屏幕时）。这个属性是在 APIlevel3 中引入的	android：noHistory＝["true"︱"false"]
android：permission	权限与安全机制解析	android：permission＝"string"
android：process	一个 Activity 运行时所在的进程名，所有程序组件运行在应用程序默认的进程中，这个进程名跟应用程序的包名一致	android：process＝"string"
android：screenOrientation	Activity 显示的模式，"unspecified"为默认值；"landscape"为风景画模式，宽度比高度大一些；"portrait"为肖像模式，高度比宽度大；"user"为用户的设置；另外还有"behind""sensor""nonsensor"	android：screenOrientation＝["unspecified"︱"user"︱"behind"︱"landscape"︱"portrait"︱"sensor"︱"nonsensor"]

续表

属　性	描　述	调 用 方 法
android：stateNotNeeded	是否 Activity 被销毁和成功重启并不保存状态	android：stateNotNeeded=["true"|"false"]
android：taskAffinity	Activity 的亲属关系，默认同一个应用程序下的 Activity 有相同的关系	android：taskAffinity="string"
android：theme	Activity 的样式主题，如果没有设置，则 Activity 的主题样式从属于应用程序，请参见 <application>元素的 theme 属性	android：theme="resourceortheme"
android：windowSoftInputMode	Activity 主窗口与软键盘的交互模式，自 APIlevel3 被引入	android：windowSoftInputMode=[one-ormoreof："stateUnspecified""state-Unchanged""stateHidden""stateAlwaysHidden""stateVisible""stateAlwaysVisible""adjustUnspecified""adjustResize""adjustPan"]>

5.3　项目运行

前面对 Activity 生命周期进行了讲解，本节将通过一个实例让大家生动地体验项目的生命周期是如何进行的，即如何在 App 项目中创建 Activity、设置 Activity，以及启动 Activity、关闭 Activity。

在 Android 开发中，在 src 文件夹下会自动生成一个 MainActivity.java 文件，其中实现了对 Activity 的定义及 onCreate()方法的调用，具体代码如下：

```
public class MainActivity extends ActionBarActivity {

    @Override
    protected void onCreate(Bundle savedInstanceState) {
        super.onCreate(savedInstanceState);
        setContentView(R.layout.activity_main);
    }
```

通过调用 onCreate()方法实现 savedInstanceState,即每次启动一个 Activity 后,能够保持在一个界面内容不变,直到启动另一个 Activity 才能修改界面内容视图。

5.3.1 创建新的 Activity

【例 5.1】 创建一个新的 Activity,并通过代码实现新建 Activity 的设置、启动。除了新建项目时自带的 Activity 以外,还需要另外新建其他 Activity,以实现其他界面内容的布局、调用与交互。为了定义新创建的 Activity,并在生命周期进行运转时告知 Android 系统,还需在本项目根目录下的 AndroidManifest.xml 文件中进行标注,具体代码如下:

```
<application
        android:label = "@string/app_name" >
    <activity
        android:label = "@string/app_name"
        android:name = ".MainActivity" >
        <intent - filter >
            <action android:name = "android.intent.action.MAIN" />
            <category android:name = "android.intent.category.LAUNCHER" />
        </intent - filter >
    </activity>
    <activity
        android:name = ".NewActivity"
        android:label = "新建 Activity"
        >
    </activity>
</application>
```

（定义创建新的Activity）

将新建的 Activity 定义为 NewActivity 的名字后,继续通过 Android Studio 创建 NewActivity 活动。具体过程如下:

步骤 1：右击 layout 文件夹,选择 New→Activity→Basic Activity 命令,如图 5-3 所示。

步骤 2：在弹出的 New Android Activity 对话框中填写新建 Activity 的相关配置,包括 Activity Name、Layout Name、Title、Package Name 等信息,如图 5-4 所示。

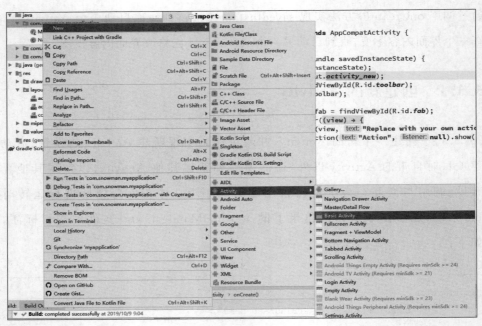

图 5-3　新建 Activity

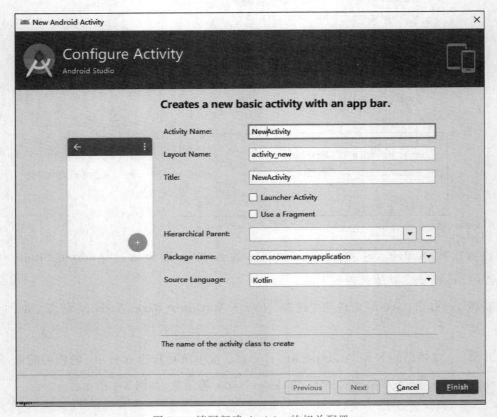

图 5-4　填写新建 Activity 的相关配置

由 Android Studio 自动生成的代码如下：

```
package com.mingrisoft;

import android.app.Activity;
import android.os.Bundle;

public class NewActivity extends Activity {

    @Override
    protected void onCreate(Bundle savedInstanceState) {
        // TODO Auto-generated method stub
        super.onCreate(savedInstanceState);
    }

}
```

步骤3：为了使新建的 NewActivity 运行时有新的界面可以调用，需要为 NewActivity 新建一个布局文件，命名为 newactivity.xml，如图 5-5 所示。然后在 NewActivity.java 文件中添加一句脚本，内容为调用 setContentView() 方法为 NewActivity 指定布局文件 newactivity.xml，具体代码如下：

图 5-5　在项目中创建新活动与新布局的结构

```
package com.mingrisoft;

import android.app.Activity;
import android.os.Bundle;

public class NewActivity extends Activity {

    @Override
    protected void onCreate(Bundle savedInstanceState) {
        // TODO Auto-generated method stub
        super.onCreate(savedInstanceState);
        setContentView(R.layout.newactivity);   ——→ 增加的脚本
    }

}
```

5.3.2 为新建 Activity 设置属性

为了避免新建 Activity 启动时抛出异常，需要在 AndroidManifest.xml 文件中对 NewActivity 进行属性设置，具体代码如下：

```
<manifest xmlns:android="http://schemas.android.com/apk/res/android"
    package="com.mingrisoft"
    android:versionCode="1"
    android:versionName="1.0" >

    <uses-sdk android:minSdkVersion="15" />

    <application
        android:icon="@drawable/ic_launcher"
        android:label="@string/app_name" >
        <activity
            android:label="@string/app_name"
            android:name=".MainActivity" >
            <intent-filter >
                <action android:name="android.intent.action.MAIN" />

                <category android:name="android.intent.category.LAUNCHER" />
            </intent-filter>
        </activity>
```

```
    <activity
        android:icon = "@drawable/ic_launcher"
        android:name = ".NewActivity"
        android:label = "新建 Activity"
        android:launchMode = "singleTask"
        android:screenOrientation = "portrait"
        android:windowSoftInputMode = "stateHidden"
        >
    </activity>
</application>
```

⟶ 为NewActivity设置属性

```
</manifest>
```

5.3.3　启动 Activity

在本项目中不止一个 Activity,因此在启动 Activity 时需要调用 startActivity()方法,其语法格式为:

```
Public void startActivity(Intent newintent)
```

其中,Intent 用于 Activity 之间的数据传递,每个 Intent 都要与一个 Activity 相对应。该方法应编写于 layout 文件夹下新建的 newactivity.xml 文件中,具体代码如下:

```
<?xml version = "1.0" encoding = "utf-8"?>
<LinearLayout xmlns:android = "http://schemas.android.com/apk/res/android"
    android:layout_width = "match_parent"
    android:layout_height = "match_parent"
    android:orientation = "vertical" >

        <TextView
        android:id = "@ + id/textView1"
        android:layout_width = "wrap_content"
        android:layout_height = "wrap_content"

        />
    Intent intent = new Intent(MainActivity.this,ToStartActivity.class);
        startActiviity(newintent);
</LinearLayout>
```

5.3.4　关闭 Activity

关闭 Activity 的方法较为简单,如果只有一个 Activity,只需调用 finish()方法即可,其

语法格式如下:

```
Public void finish()
```

但如果有多个 Activity,则需要调用 finishActivity()方法来指定关闭 Activity 对象,SDK 的官方说明文档如下:

```
• public void finishActivity (int requestCode)
• Since: API Level 1
• Force finish another activity that you had previously started with startActivityForResult(Intent, int).
• Parameters requestCode  The request code of the activity that you had given to startActivityForResult(). If there are multiple activities started with this request code, they will all be finished.
```

```
package com.mingrisoft;

import android.app.Activity;
import android.os.Bundle;

public class NewActivity extends Activity {

    @Override
    protected void onCreate(Bundle savedInstanceState) {
        // TODO Auto-generated method stub
        super.onCreate(savedInstanceState);
        setContentView(R.layout.newactivity);
        Button btn1 = (Button)findViewById(R.id.btn1);
        btn1.setOnClickListener(new OnClickListener() {

    @Override
        public void onClick(View v) {
        ActivityA.this.finishActivity(1);

        }
    });
}
```

5.4 项目结案

本项目通过实例的方式从 Activity(活动类)的 4 种基本状态谈起,再梳理 Activity 的生命周期,并在此基础上演示了如何创建新的 Activity、设置 Activity 以及控制 Activity 状态等操作。总结本项目内容,需要大家对以下 3 个部分有较为深刻的理解:

（1）创建新 Activity 的流程。

（2）启动 Activity 与关闭 Activity 的方法。

（3）Activity 的生命周期运转方式。

5.5 项目练习

1. 开发一款 App，并在其中设置两个 Activity，调用 startActivity（）方法实现两个 Activity 的启动。

2. 开发一款 App，并在其中创建 3 个 Activity，调用 finishActivity（）方法实现依次结束 3 个 Activity。

3. 请用一款 App 的程序代码，对照 Activity 生命周期示意图，详细解析 App 的生命周期运转情况。

项目6

设置App的UI组件

6.1 项目目标：添加与设置 App 的 UI 组件

对于 App 开发与设计，最为关键的是用户的信息传递以及与用户之间的交互。在用 Android 开发 App 时，最常用到的与用户发生联系的内容是 App 界面上各种不同类型的组件。可以说，组件是构成 App 界面的关键要素，也是实现人机交互的重要元素。本项目将通过大量的实例为大家讲解 App 的组件，包括基本组件和高级组件的各种类型及应用。希望大家通过本项目的学习，能够实现交互界面的设计与开发。

6.2 项目准备

6.2.1 介绍 UI 组件：TextView 及其子类

TextView 意为文本框，它在 App 的 UI 界面开发中应用十分广泛。TextView 继承了

View 类，而 CheckedTextView、EditText、Chronometer、Button、TextClock 等类继承了 TextView，如图 6-1 所示。

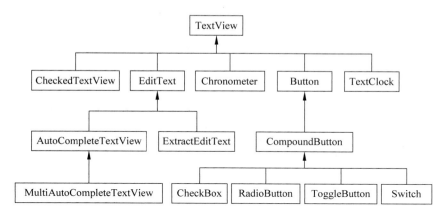

图 6-1　TextView 及其子类继承关系结构图

1. TextView

Android 中的 TextView(文本框)组件用来存放文本信息，并将此信息呈现在手机界面上。相较于 Java 开发中的 TextView，它的包容性更强，不仅能够存放多行文本信息，还能够兼容带图片的文本信息。

在界面上添加 TextView 组件，可以用以下语法通过< TextView ></ TextView >标记来编写，具体代码如下：

```
< TextView
属性列表
>
</TextView >
```

在 TextView 组件中包含大量支持 XML 的属性，并通过调用相关方法对显示的文本进行设置，如表 6-1 所示。

表 6-1　TextView 支持的 XML 属性与相关方法

XML 属性	相 关 方 法	说　　明
android：autoLink	setAutoLinkMask(int)	是否将符合指定格式的文本转换为可单击的超链接形式
android：autoText	setKeyLinstener(KeyLinstener)	控制是否将 URL、Email 等地址自动转换为可单击的链接

续表

XML 属性	相 关 方 法	说 明
android：linksClickable	setLinksClickable(boolean)	控制该文本框的 URL、Email 等链接是否可以单击
android：capitalize	setKeyLinstener(KeyLinstener)	控制是否将用户输入的文本转换为大写字母
android：cursorVisible	setCursorVisible(boolean)	设置该文本框的光标是否可见
android：drawableBottom	setCompoundDrawablesWithIntrinsicBounds(Drawable,Drawable,Drawable,Drawable)	在文本框底端绘制图像
android：drawableTop	setCompoundDrawablesWithIntrinsicBounds(Drawable,Drawable,Drawable,Drawable)	在文本框顶端绘制图像
android：drawableEnd	setCompoundDrawablesWithIntrinsicBounds(Drawable,Drawable,Drawable,Drawable)	在文本框结尾绘制图像
android：drawableLeft	setCompoundDrawablesWithIntrinsicBounds(Drawable,Drawable,Drawable,Drawable)	在文本框左边绘制图像
android：drawableRight	setCompoundDrawablesWithIntrinsicBounds(Drawable,Drawable,Drawable,Drawable)	在文本框右边绘制图像
android：drawablePadding	setCompoundDrawablesWithIntrinsicBounds(Drawable,Drawable,Drawable,Drawable)	设置文本框内文本与图形的间距
android：drawableStart	setCompoundDrawablesWithIntrinsicBounds(Drawable,Drawable,Drawable,Drawable)	在文本框开始绘制图像
android：ellipsize	setEllipsize(TextUitls,TruncateAt)	当显示文本超过了 TextView 的宽度时如何处理文本
android：ems	setEms(int)	设置文本框的宽度
android：height	setHeight(int)	设置文本框的高度
android：fontFamily	setTypeface(Typeface)	设置文本框内文本的字体
android：gravity	setGravity(int)	设置文本框内文本的对齐方式
android：hint	setHint(int)	设置文本框内容为空时，文本框内默认显示的提示文字
android：inputType	setRawInputType(int)	设置文本框输入方式
android：lines	setLines(int)	设置文本框默认占几行
android：maxEms	setMaxEms(int)	设置文本框最大宽度
android：maxHeight	setMaxHeight(int)	设置文本框最大高度
android：maxLength	setMaxLength(int)	设置文本框最大字符长度

续表

XML 属性	相关方法	说 明
android：maxWidth	setMaxWidth(int)	设置文本框最大宽度
android：password	setTransformatinMethod(TransformatinMethod)	设置文本框是一个密码框
android：textColor	setTextColor(ColorStateList)	设置文本框中文本的颜色
android：textColorLink	setTextColorLink(int)	设置文本框中链接的颜色
android：textSize	setTextSize(float)	设置文本框的字体大小
android：textStyle	setTextStyle(Typeface)	设置文本框内的字体风格,例如粗体、斜体等

2. EditText

EditText 意为编辑文本,它与 TextView 组件之间的区别在于,EditText 组件能够实现用户对文本的输入功能,而 TextView 不能。对于 EditText 组件而言,有多种形式的文本类型可以被用户输入界面,包括单行文本、多行文本及特定文本形式(电话号码、密码、电子邮件地址等)。

在界面上添加 EditText 组件,可以用以下语法通过<EditText></EditText>标记来编写,具体代码如下：

```
<EditText>
属性列表
>
</EditText>
```

为 EditText 组件的输入类型赋值有多种选择,具体类型与调用方法如表 6-2 所示。

表 6-2 EditText 组件的输入类型及调用方法

输入类型	调用方法	含 义
textCapSentences	android：inputType="textCapSentences"	仅第一个字母大写
textAutoCorrect	android：inputType="textAutoCorrect"	文本自动修正
textAutoComplete	android：inputType="textAutoComplete"	文本自动完成
textMultiLine	android：inputType="textMultiLine"	多行输入
textImeMultiLine	android：inputType="textImeMultiLine"	输入法多行
textNoSuggestions	android：inputType="textNoSuggestions"	不提示
textUrl	android：inputType="textUrl"	URL 格式

续表

输 入 类 型	调 用 方 法	含 义
textEmailAddress	android：inputType="textEmailAddress"	电子邮件地址格式
textEmailSubject	android：inputType="textEmailSubject"	邮件主题格式
textShortMessage	android：inputType="textShortMessage"	短消息格式
textLongMessage	android：inputType="textLongMessage"	长消息格式
textPersonName	android：inputType="textPersonName"	人名格式
textPostalAddress	android：inputType="textPostalAddress"	邮政格式
textPassword	android：inputType="textPassword"	密码格式
textVisiblePassword	android：inputType="textVisiblePassword"	密码可见格式
textWebEditText	android：inputType="textWebEditText"	作为网页表单的文本格式
textFilter	android：inputType="textFilter"	文本筛选格式
textPhonetic	android：inputType="textPhonetic"	拼音输入格式
number	android：inputType="number"	数字格式
numberSigned	android：inputType="numberSigned"	有符号数字格式
numberDecimal	android：inputType="numberDecimal"	可以带小数点的浮点格式
phone	android：inputType="phone"	拨号键盘
datetime	android：inputType="datetime"	日期键盘
time	android：inputType="time"	时间键盘

3. Chronometer

Chronometer 意为计时器，用于记录时间的长短，并以文本的形式展现在界面上。一般而言，Chronometer 组件常用的实现方法如表 6-3 所示。

表 6-3 Chronometer 组件的方法

方 法 名	说 明
setFormat()	设置计时时间的显示样式
setBase()	设置计时器的起始时间
Start()	设置计时器开始
Stop()	设置计时器停止
setOnChronometerTickListener()	监听计时器事件

4. Button

在 Android 开发中,按钮分为单选按钮(RadioButton)和按钮组(RadioGroup)。在默认情况下,单选按钮以圆形的图标显示,在设计时可将按钮的说明文字放在旁边。当把多个单选按钮放置在一个按钮组中时,按钮之间可以实现非此即彼的选择状态,即当选中其中某个单选按钮时,按钮组中的其他按钮将自动取消被选中。RadioButton 是 Button 的子类,可以直接使用其支持的各种属性,并通过在 XML 布局文件中使用< RadioButton >标记的方法实现在屏幕中添加单选按钮,基本格式如下:

```
< RadioButton
    android:id = "@ + id/ID 号"
    android:text = "显示文本"
    android:layout_width = "wrap_content"
    android:layout_height = "wrap_content"
    android:checked = "true|false"
    >
</RadioButton >
```

一般而言,RadioButton 组件需要与 RadioGroup 组件联合使用,即多个单选按钮组成一个单选按钮组,并在 XML 布局文件中添加 RadioGroup 组件,基本格式如下:

```
< RadioGroup
    android:id = "@ + id/ID 号"
    android:layout_width = "wrap_content"
    android:layout_height = "wrap_content"
    android:orientation = "horizontal"
    >
    <!-- 添加多个 RadioGroup 组件 -->
</RadioGroup >
```

6.2.2 介绍 UI 组件:ImageView 及其子类

在手机界面中,ImageView(图像视图)组件是十分直观的,它也是设计中被关注的重点。顾名思义,ImageView 是显示图形图像的功能性组件。一般而言,在 Android 中可以通过两种方式实现在手机界面上添加图像视图:一种方式是在 XML 布局文件中添加< ImageView ></ ImageView >标记;另一种方式是在 Java 中用 new 关键字来创建。在

Android 编程中常用第一种方式,即将放置在 res 下 drawable 文件夹中的图像在 XML 布局文件中进行调用,具体代码如下:

```
<ImageView
    属性列表
>
</ImageView>
```

ImageView 组件派生出另外两个类:一个是 ImageButton,即图片按钮;另一个是 QuickContactBadge,即用于显示关联到特定联系人的图片。ImageView 组件及其派生类支持的常用 XML 属性如表 6-4 所示。

表 6-4　ImageView 支持的 XML 属性和相关的方法

XML 属性	相 关 方 法	说　　明
android:adjustViewBounds	setAdjustViewBounds(boolean)	是否调整边界来保持所显示图片的长宽比
android:baseline	setBaseline(int)	设置视图内基线的偏移量
android:baselineAlignBottom	setBaselineAlignBottom(boolean)	如果设置为 true,将父视图基线与 ImageView 底部边缘对齐
android:cropToPadding	setCropToPadding(boolean)	如果设置为 true,组件将会被裁剪到保留 ImageView 的 padding
android:maxHeight	setMaxHeight(int)	设置最大高度
android:maxWidth	setMaxWidth(int)	设置最大宽度
android:scaleType	setScaleType(ImageView.ScaleType)	设置显示图片如何缩放和移动以适应 ImageView 的大小
android:src	setImageResource(int)	设置所显示的 Drawable 对象的 ID
android:tint	setColorFilter(int,PorterDuff.Mode)	将图片渲染成指定的颜色

6.2.3　介绍 UI 组件:AdapterView 及其子类

AdapterView 组件是一个重要的视图组件,继承了 ViewGroup,它本身是一个抽象类,AdapterView 及其子类的继承关系如图 6-2 所示。其派生的子类在用法上十分相似,只是显示的界面有一定的区别。

AdapterView 派生了 3 个类,分别是 AbsListView、AbsSpinner 和 AdapterViewAnimator,

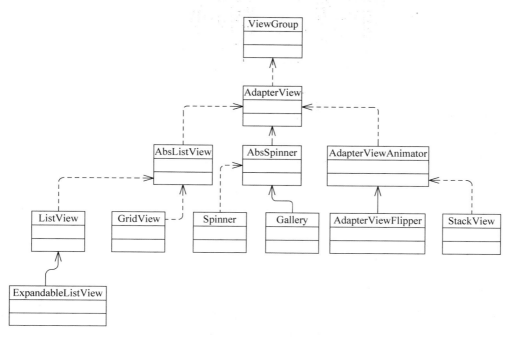

图 6-2　AdapterView 及其子类的继承关系示意图

它们都可以实现以列表的形式显示数据。其中，ListView、Gallery 和 GridView 等是上述类的子类，能够以多种形式显示视图效果。值得注意的是，ListView、GridView、Spinner、Gallery 等 AdapterView 只是容器，而 Adapter 负责提供每个"列表项"组件，AdapterView 负责采用合适的方式显示这些列表项。

1. ListView

ListView（列表视图）在 AdapterView 的所有子类中最常用，它以垂直的列表形式显示列表数据信息。创建 ListView 有两种方法：一是直接使用 ListView 组件创建；二是让 Activity 继承 ListActivity。使用 ListView 组件创建，通常可以在 XML 布局文件中添加 <ListView></ListView>标记，或者在 Java 文件中通过 new 关键字创建。在 XML 布局文件中添加 ListView 的基本格式如下：

```
<ListView
属性列表
>
</ListView>
```

ListView 的 XML 属性如表 6-5 所示。

表 6-5 ListView 的 XML 属性

XML 属性	说 明
android：divider	在列表条目之间显示的 drawable 或 color
android：dividerHeight	用来指定 divider 的高度
android：entries	构成 ListView 的数组资源的引用。对于某些固定的资源，这个属性提供了比在程序中添加资源更加简便的方式
android：footerDividersEnabled	当设为 false 时，ListView 将不会在各个 footer 之间绘制 divider，其默认为 true
android：headerDividersEnabled	当设为 false 时，ListView 将不会在各个 header 之间绘制 divider，其默认为 true

2. Spinner

Spinner 组件即列表选择框，能够实现弹出相应菜单供用户选择的视图效果。创建 Spinner 有两种方法：一是在 XML 布局文件中添加< Spinner ></Spinner >标记；二是在 Java 文件中通过 new 关键字进行创建。在 XML 布局文件中添加 Spinner 的基本格式如下：

```
< LinearLayout
        android:layout_width = "fill_parent"
        android:layout_height = "fill_parent"
        android:orientation = "vertical" >

    < Spinner
            android:id = "@ + id/spinner1"
            android:layout_width = "wrap_content"
            android:layout_height = "wrap_content"
            android:entries = "@array/languages"
    >
    </Spinner >
</LinearLayout >
```

Spinner 的 XML 属性如表 6-6 所示。

表 6-6 Spinner 的 XML 属性

XML 属性	相 关 方 法	说 明
android：entries		使用数组资源设置该下拉列表框的列表项目
android：dropDownHorizontalOffset	setdropDownHorizontalOffset(int)	设置下拉列表框的水平偏移距离

续表

XML 属性	相关方法	说明
android：dropDownVerticalOffset	setdropDownVerticalOffset(int)	设置下拉列表框的垂直偏移距离
android：dropDownWidth	setdropDownWidth(int)	设置下拉列表框的宽度
android：popupBackground	setpopupBackgroundResource(int)	设置下拉列表框的背景颜色
android：prompt		设置下拉列表框的提示信息

3. GridView

GridView 即为网格视图，用于在界面上实现按行、列分布的方式显示多个组件。GridView 和 ListView 具有相同的父类 AbsListView，因此 GridView 和 ListView 很相似，但是 ListView 只显示一列，而 GridView 可以显示多列。创建 GridView 一般通过在 XML 布局文件中添加< GridView ></GridView >标记来实现。在 XML 布局文件中添加 GridView 的基本格式如下：

```
< GridView
    属性列表
>
</GridView >
```

GridView 常用的 XML 属性如表 6-7 所示。

表 6-7　GridView 常用的 XML 属性

XML 属性	相关方法	说明
android：numColumns	setnumColumns(int)	设置列数
android：columnWidth	setcolumnWidth(int)	设置列的宽度
android：gravity	setGravity(int)	设置对齐方式
android：verticalSpacing	setverticalSpacing(int)	设置两行之间的边距
android：stretchMode	setstretchMode(int)	设置缩放与列宽大小同步
android：scrollbars		隐藏 GridView 的滚动条

6.2.4　介绍 UI 组件：ProgressBar 及其子类

ProgressBar 即为进度条组件，它继承自 View 类，直接子类有 AbsSeekBar 和 ContentLoadingProgressBar，间接子类有 RatingBar 和 SeekBar（这两个组件也是最常使用

的子类)。它们的使用方法类似,仅在显示界面上有一定的区别。在 API 文档中对 ProgressBar 的说明如图 6-3 所示。

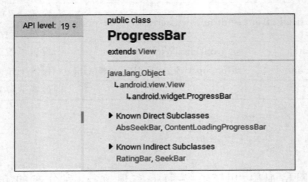

图 6-3 API 文档对 ProgressBar 的说明

1. ProgressBar

ProgressBar(进度条)一般通过颜色来表示进度值,创建的方法是在 XML 布局文件中添加< ProgressBar ></ProgressBar >标记来实现。在 XML 布局文件中添加 ProgressBar 的基本格式如下:

```
< ProgressBar
    属性列表
>
</ProgressBar >
```

ProgressBar 常用的 XML 属性如表 6-8 所示。

表 6-8 ProgressBar 常用的 XML 属性

XML 属性	说 明
android：animationResolution	动画超时时间,必须是整数值,例如"100"
android：indeterminate	该属性设置为 true,表示不精确显示进度
android：indeterminateBehavior	设置当选择不精确显示进度时如何描述到达最大值
android：indeterminateDrawable	设置当选择不精确显示进度时所绘制的 Drawable 对象
android：indeterminateDuration	设置不精确显示进度的持续时间
android：indeterminateOnly	设置只采用不精确显示进度模式(状态保持模式将不会工作)
android：interpolator	设置动画的速度
android：max	设置进度可以达到的最大值

续表

XML 属性	说　　明
android：maxHeight	可选参数，设置 View 的最大高度
android：maxWidth	可选参数，设置 View 的最大宽度
android：minHeight	可选参数，设置 View 的最小高度
android：minWidth	可选参数，设置 View 的最小宽度
android：mirrorForRtl	定义是否需要反映在 RTL 模式的相关画板，默认为 false
android：progress	设置该进度已完成的进度值
android：progressDrawable	设置该进度条轨道对应的 Drawable 对象
android：secondaryProgress	二级进度条，主要用于缓存使用的场景

ProgressBar 支持多种风格，可以直接通过 style 属性来设置，在 API 中也给出了如表 6-9 所示的属性。

表 6-9　ProgressBar 常用的 style 属性

style 属性	说　　明
Widget. ProgressBar. Horizontal	水平进度条
Widget. ProgressBar. Small	小环进度条
Widget. ProgressBar. Large	大环进度条
Widget. ProgressBar. Inverse	普通大小的环形进度条
Widget. ProgressBar. Small. Inverse	小环形进度条
Widget. ProgressBar. Large. Inverse	大环形进度条

另外一种方式就是使用系统的 att 属性，例如：

(1) style＝"？android：attr/progressBarStyle"

(2) style＝"？android：attr/progressBarStyleHorizontal"

(3) style＝"？android：attr/progressBarStyleInverse"

(4) style＝"？android：attr/progressBarStyleLarge"

(5) style＝"？android：attr/progressBarStyleLargeInverse"

(6) style＝"？android：attr/progressBarStyleSmall"

(7) style＝"？android：attr/progressBarStyleSmallInverse"

(8) style＝"？android：attr/progressBarStyleSmallTitle"

2. SeekBar

SeekBar 即为滑动条,它通过滑块来表示进度数值,而且允许用户通过拖动滑块来改变数值。其具体形式与 Windows 的音量调节器类如图 6-4 所示。

图 6-4 Windows 的音量调节器类

SeekBar 的创建通常是在 XML 布局文件中添加 <SeekBar></SeekBar> 标记来实现。在 XML 布局文件中添加 SeekBar 的基本格式如下:

```
<SeekBar
        android:id = "@ + id/seekbar"
        android:layout_width = "579dp"
        android:layout_height = "wrap_content"
        android:layout_centerVertical = "true"
        android:layout_marginLeft = "100dp"
        android:progressDrawable = "@drawable/seekbar_style"
        android:minHeight = "10dp"
        android:maxHeight = "10dp"
        android:thumbOffset = "1.0dp"
        android:thumb = "@drawable/seekbar_point_style" >
</SeekBar>
```

3. RatingBar

RatingBar 即为星级评分条,可以通过对星星的选择情况来表现进度。一般而言,星级评分条常用于用户评价、客户满意度等数据的获取。RatingBar 的创建通常是在 XML 布局文件中添加 <RatingBar></RatingBar> 标记来实现。在 XML 布局文件中添加 RatingBar 的基本格式如下:

```
<RatingBar
    属性列表
>
</RatingBar>
```

RatingBar 常用的 XML 属性如表 6-10 所示。

表 6-10 RatingBar 常用的 XML 属性

XML 属性	方 法	说 明
android:isIndicator		设置该星级评分是否允许用户改变(true 为不允许改变)
android:numStars	getProgress()	设置该星级评分条总共有多少个星星

续表

XML 属性	方法	说明
android：rating	getRating()	设置该星级评分条默认的星级
android：stepSize	getstepSize()	设置每次最少需要改变多少个星级

6.2.5 介绍 UI 组件：ViewAnimator 及其子类

ViewAnimator 是一个基类，继承了 FrameLayout 父类，因此它同样表现出帧布局的特征，即可以将多个 View 组件叠在一起，它所额外增加的功能是在 View 切换时表现出动画效果。它的子类有 ViewSwitcher、ViewFlipper、ImageSwitcher、TextSwitcher。ViewAnimator 常用的 XML 属性如表 6-11 所示。

表 6-11 ViewAnimator 常用的 XML 属性

XML 属性	说明
android：animateFirstView	设置 ViewAnimator 显示第一个 View 组件时是否显示动画
android：inAnimation	设置 ViewAnimator 显示组件时所使用的动画
android：outAnimation	设置 ViewAnimator 隐藏组件时所使用的动画

1. ImageSwitcher

ImageSwitcher(图像切换器)组件的主要功能是完成图片的切换显示。例如用户在进行图片浏览时，可以通过单击按钮逐张切换显示的图片，在进行切换时还可以加入一些动画效果。ImageSwitcher 的创建通常是在 XML 布局文件中添加＜ImageSwitcher＞＜/ImageSwitcher＞标记来实现。在 XML 布局文件中添加 ImageSwitcher 的基本格式如下：

```
＜ImageSwitcher
    属性列表
＞
＜/ImageSwitcher＞
```

2. TextSwitcher

TextSwitcher 即为文字切换器，它是 ViewSwitcher 的子类。从 ViewSwitcher 来看，是 View 交换器，TextSwitcher 继承自 ViewSwitcher，显然是交换 TextView。TextSwitcher

的创建通常是在 XML 布局文件中添加< TextSwitcher ></TextSwitcher >标记来实现。在 XML 布局文件中添加 TextSwitcher 的基本格式如下：

```
< TextSwitcher
    属性列表
>
</TextSwitcher >
```

6.3 项目运行

6.3.1 在 UI 中设计文本框：TextView 组件实例

【例 6.1】 设计一款 App，其界面由 TextView 组件组成，对 TextView 组件中的信息进行如下设置：文字颜色由红色和黄色组成，并为文字信息添加一个图片信息，设定不同的文字尺寸。

视频讲解

步骤 1：新建名为 6.1 的 Android 项目文件。选择 File|New|Android Application Project 命令，在弹出的 New Android Application 对话框中的 Application Name 文本框中输入"6.1"，再单击 Finish 按钮。

步骤 2：设置布局文件 main.xml。单击 6.1 项目文件夹，然后单击 res 下 layout 文件夹中的 main.xml 文件，在该布局文件中设置两个 TextView 组件，其中一个组件实现文字配图片（文字为红色，尺寸为 25sp），另一个组件实现多行文本信息（文字为黄色，尺寸为 15sp）。其具体代码如下：

```xml
<?xml version = "1.0" encoding = "utf-8"?>
<LinearLayout                                              线性布局
    xmlns:android = "http://schemas.android.com/apk/res/android"
    android:layout_width = "fill_parent"
    android:layout_height = "fill_parent"
    android:background = "@drawable/background02"
    android:orientation = "vertical" >

    <TextView                                              第一个TextView组件
        android:id = "@ + id/textView1"
```

```
            android:layout_width = "300dp"
            android:layout_height = "wrap_content"
            android:layout_gravity = "center"
            android:drawableTop = "@drawable/icon"        ——→ 添加配图
            android:gravity = "center"
            android:text = "@string/message1"             ——→ 添加文字信息,详见string.xml
            android:textColor = "#F00"                    ——→ 设置文字颜色为红色
            android:textSize = "25sp" />                  ——→ 设置文字尺寸

        <TextView
            android:id = "@+id/textView2"
            android:layout_width = "wrap_content"
            android:layout_height = "0dp"
            android:layout_gravity = "center_horizontal"
            android:layout_weight = "1"
            android:text = "@string/message2"
            android:textColor = "#FF0"
            android:textSize = "15sp"
            android:width = "300dp" />

</LinearLayout>
```

步骤3:设置文本信息。单击6.1项目文件夹,然后单击res下values文件夹中的strings.xml文件,在<resources></resources>标记之间添加文本内容。其具体代码如下:

```
<?xml version = "1.0" encoding = "utf-8"?>
<resources>
    <string name = "hello">再别康桥</string>
    <string name = "app_name">6.3.1 TextView组件实例</string>
    <string name = "message1">再别康桥</string>
    <string name = "message2">轻轻的我走了,正如我轻轻的来;我轻轻的招手,作别西天的云彩。……那河畔的金柳,是夕阳中的新娘;波光里的艳影,在我的心头荡漾。……软泥上的青荇,油油的在水底招摇;在康河的柔波里,我甘心做一条水草!……

</string>
</resources>
```

步骤4:运行程序,效果如图6-5所示。

6.3.2 在UI中设计可编辑文本框:EditText组件实例

【例6.2】 设计一款运动类App,使用EditText组件设计其"用户信息"页面的内容。

步骤1:新建一个名为6.2的Android项目,打开res下layout文件夹中的main.xml文件,删除原有默认布局代码脚本,设置线性布局,并在该布局中设置背景图片,添加3个

图 6-5 TextView 组件实例的运行效果

EditText 类,引导用户填写用户名、电话号码和电子邮箱信息。其具体代码如下:

```xml
<?xml version = "1.0" encoding = "utf-8"?>
< LinearLayout xmlns:android = "http://schemas.android.com/apk/res/android"
    android:id = "@ + id/tableLayout1"
    android:layout_width = "fill_parent"
    android:layout_height = "fill_parent"
    android:background = "@drawable/bg"
    android:orientation = "horizontal" >

    < TextView
        android:layout_width = "wrap_content"
        android:layout_height = "wrap_content"
        android:text = "@string/message1"
        android:height = "60dp" />
    < EditText android:id = "@ + id/nickname"
        android:hint = "@string/message2"
        android:inputType = "textEmailAddress"
        android:layout_width = "300dp"
        android:layout_height = "wrap_content"
        android:singleLine = "true"
        />
```

实现TextView+EditText
健身达人昵称 请输入达人昵称

```
<TextView
    android:layout_width = "wrap_content"
    android:layout_height = "wrap_content"
    android:text = "@string/message3"
    android:height = "60dp" />
<EditText android:id = "@+id/phonenumber"
    android:layout_width = "300dp"
    android:inputType = "number"
    android:hint = "@string/message4"
    android:layout_height = "wrap_content"
    />
```

实现TextView+EditText
健身达人手机号 请输入达人手机号

```
<TextView
    android:layout_width = "wrap_content"
    android:layout_height = "wrap_content"
    android:text = "@string/message5"
    android:height = "60dp" />
<EditText android:id = "@+id/emailaddress"
    android:layout_width = "300dp"
    android:hint = "@string/message6"
    android:layout_height = "wrap_content"
    android:inputType = "textEmailAddress"
    />
```

实现TextView+EditText
健身达人邮箱 请输入邮箱地址

```
</LinearLayout>
```

步骤2：在res下values文件夹中的strings.xml文件中为字符串编写具体内容。其具体代码如下：

```
<?xml version = "1.0" encoding = "utf-8" standalone = "no"?>
<resources>
    <string name = "hello">EditText组件实例</string>
    <string name = "app_name">6.2  EditText组件实例</string>
    <string name = "message1">健身达人昵称</string>
    <string name = "message2">请输入达人昵称</string>
    <string name = "message3">健身达人手机号</string>
    <string name = "message4">请输入达人手机号</string>
    <string name = "message5">健身达人邮箱</string>
    <string name = "message6">请输入邮箱地址</string>
</resources>
```

通过Android本地化文本编辑器窗口能够看到字符串的具体信息，如图6-6所示。

步骤3：获取用户输入到EditText中的信息。通过在布局文件main.xml中继续定义按钮，实现在UI界面上获取信息的设置。添加方式为在<LinearLayout></LinearLayout>之间添加如下代码：

app_name	app\src\main\res	☐	4.2
hello	app\src\main\res	☐	EditText 组件实例
message1	app\src\main\res	☐	健身达人昵称
message2	app\src\main\res	☐	请输入达人昵称
message3	app\src\main\res	☐	健身达人手机号
message4	app\src\main\res	☐	请输入达人手机号
message5	app\src\main\res	☐	健身达人邮箱
message6	app\src\main\res	☐	请输入邮箱地址

图 6-6　Android 本地化文本编辑器窗口

```
<Button android:text = "@string/message7"
        android:id = "@ + id/button1"
        android:layout_width = "wrap_content"
        android:layout_height = "wrap_content"/>
<Button android:text = "@string/message8"
        android:id = "@ + id/button2"
        android:layout_width = "wrap_content"
        android:layout_height = "wrap_content"/>
```

步骤 4：为按钮上的字符串编写具体内容。在 res 下 values 文件夹中的 strings.xml 文件中添加如下代码：

```
<string name = "message8">重置</string>
<string name = "message7">注册</string>
```

步骤 5：为按钮设置监听事件，用于获取用户在 EditText 组件中输入的文本信息。该段代码应在 src 下的 MainActivity.java 文件中进行编写，具体代码如下：

```
package com.mingrisoft;

import android.app.Activity;
import android.os.Bundle;
import android.util.Log;
import android.view.View;
import android.view.View.OnClickListener;
import android.widget.Button;
import android.widget.EditText;
```

```
public class MainActivity extends Activity {
    /** Called when the activity is first created. */
    @Override
    public void onCreate(Bundle savedInstanceState) {
        super.onCreate(savedInstanceState);
        setContentView(R.layout.main);
        Button button1 = (Button)findViewById(R.id.button1);
        button1.setOnClickListener(new OnClickListener() {

    @Override
            public void onClick(View v) {
                EditText nicknameET = (EditText)findViewById(R.id.nickname);
                String nickname = nicknameET.getText().toString();
                EditText phonenumberET = (EditText)findViewById(R.id.phonenumber);
                String phonenumber = phonenumberET.getText().toString();
                EditText emailaddressET = (EditText)findViewById(R.id.emailaddress);
                String emailaddress = emailaddressET.getText().toString();
                    Log.i("EditText 组件实例","会员昵称:" + nickname);
                    Log.i("EditText 组件实例","密码:" + phonenumber);
                    Log.i("EditText 组件实例","E-mail 地址:" + emailaddress);
                }
            });
    }
}
```

步骤 6：运行程序，得到健身 App 的用户信息界面，如图 6-7 所示。

图 6-7　EditText 组件实例运行效果图

6.3.3 在 UI 中设计计时器:Chronometer 组件实例

【例 6.3】 设计一款计时 App,使用 Chronometer 组件设计其计时器页面的内容。

步骤 1:新建一个名为 6.3 的 Android 项目,打开 res 下 layout 文件夹中的 main.xml 文件,删除原有默认布局代码脚本,设置相对布局,并在该布局中设置背景图片,添加一个 Chronometer 类。其具体代码如下:

```xml
< RelativeLayout xmlns:android = "http://schemas.android.com/apk/res/android"
    xmlns:tools = "http://schemas.android.com/tools"
    android:layout_width = "match_parent"
    android:layout_height = "match_parent"
    android:paddingBottom = "@dimen/activity_vertical_margin"
    android:paddingLeft = "@dimen/activity_horizontal_margin"
    android:paddingRight = "@dimen/activity_horizontal_margin"
    android:paddingTop = "@dimen/activity_vertical_margin"
    tools:context = "com.example.one.MainActivity"
    android:background = "@drawable/bg">

    < TextView
        android:id = "@ + id/textView1"
        android:layout_width = "wrap_content"
        android:layout_height = "wrap_content"
        android:text = "@string/hello_world"
        android:textColor = "#FFF"
        android:textSize = "30sp"/>

    < Chronometer
        android:id = "@ + id/chronometer"
        android:layout_width = "wrap_content"
        android:layout_height = "70sp"
        android:layout_alignParentBottom = "true"
        android:layout_centerHorizontal = "true"
        android:layout_marginBottom = "166dp"
        android:textColor = "#F00"
        android:textSize = "40sp" />
</RelativeLayout>
```

步骤 2:在 main.xml 文件中设置 3 个按钮,分别为"开始""停止"和"重置"。其具体代码如下:

```xml
    < Button
```

```xml
        android:id = "@+id/button1"
        android:layout_width = "wrap_content"
        android:layout_height = "wrap_content"
        android:layout_above = "@+id/chronometer"
        android:layout_alignLeft = "@+id/textView1"
        android:layout_marginBottom = "36dp"
        android:layout_marginLeft = "16dp"
        android:onClick = "start"
        android:text = "@string/message1"
        android:textColor = "#FF0" />

    <Button
        android:id = "@+id/button3"
        android:layout_width = "wrap_content"
        android:layout_height = "wrap_content"
        android:layout_alignBaseline = "@+id/button1"
        android:layout_alignBottom = "@+id/button1"
        android:layout_centerHorizontal = "true"
        android:onClick = "stop"
        android:text = "@string/message2"
        android:textColor = "#FF0"/>

    <Button
        android:id = "@+id/button2"
        android:layout_width = "wrap_content"
        android:layout_height = "wrap_content"
        android:layout_alignBaseline = "@+id/button3"
        android:layout_alignBottom = "@+id/button3"
        android:layout_marginLeft = "26dp"
        android:layout_toRightOf = "@+id/button3"
        android:onClick = "reset"
        android:text = "@string/message3"
        android:textColor = "#FF0"/>
```

步骤3：在res下values文件夹中的strings.xml文件中为App界面和按钮中的字符串编写具体内容。其具体代码如下：

```xml
<?xml version = "1.0" encoding = "utf-8" standalone = "no"?>
<resources>
    <string name = "app_name">6.3Chronometer组件实例</string>
    <string name = "hello_world">计时器界面</string>
    <string name = "action_settings">Settings</string>
    <string name = "message1">开始</string>
    <string name = "message2">停止</string>
    <string name = "message3">重置</string>
</resources>
```

步骤4：为计时器设置监听事件，用于获取用户在Chronometer组件中的计时信息。设置监听事件可以采用onClick()方法来实现。该段代码应在src下的MainActivity.java文件中进行编写，具体代码如下：

```java
package com.example.one;
import android.support.v7.app.ActionBarActivity;
import android.os.Bundle;
import android.os.SystemClock;
import android.view.View;
import android.widget.Chronometer;
import android.widget.Toast;

public class MainActivity extends ActionBarActivity {

    long getBase() {
        return 0;
    }
    String getFormat() {
        return null;
    }
    void setBase(long base) {
    }
    void setFormat(String format) {
    }

    private Chronometer chronometer;
    private long recordTime;        //记录下来的总时间
    @Override
    protected void onCreate(Bundle savedInstanceState) {
        super.onCreate(savedInstanceState);
        setContentView(R.layout.main);
        chronometer = (Chronometer) findViewById(R.id.chronometer);
        chronometer.setFormat("计时:% s");
        Toast.makeText(MainActivity.this, "" + recordTime, Toast.LENGTH_SHORT).show();
    }
    public void start(View view){
        chronometer.setBase(SystemClock.elapsedRealtime() - recordTime);
        chronometer.start();
    }
    public void stop(View view){
        chronometer.stop();
        recordTime = SystemClock.elapsedRealtime() - chronometer.getBase();
    }
    public void reset(View view){
        recordTime = 0;              //重置时间
```

```
        chronometer.setBase(SystemClock.elapsedRealtime());
    }
}
```

步骤5：运行结果演示。分别测试计时开始、停止和重置状态下App的UI界面显示情况，如图6-8所示。

图6-8　Chronometer组件的初始界面、开始计时(重置)界面、停止界面运行效果

6.3.4　在UI中设计单选按钮：RadioGroup组件实例

【例6.4】　在Android Studio中创建一个App项目，在屏幕上添加一个单选按钮组，选项内容为成绩合格或不合格。

视频讲解

步骤1：在Android Studio中创建一个名为6.4的App项目，并在项目文件中找到res下layout文件夹中的main.xml文件。

步骤2：对main.xml文件进行编写。为了单选按钮的效果在手机界面上看起来更美观，将新建项目中默认的垂直布局改为水平布局的方式。

步骤3：在水平布局管理器中添加TextView组件，并在其中设置选项说明信息，即"成绩："。

步骤4：在水平布局管理器中添加单选按钮组件(RadioGroup)，其中包含两个单选按钮(RadioButton)，并在其中设置选项说明信息，即"合格"和"不合格"。

步骤 5：在水平布局管理器中添加"提交"按钮,并设置其布局方式,如图 6-9 所示。

图 6-9　添加单选按钮组的效果图

main.xml 文件中的具体代码如下：

```xml
<?xml version = "1.0" encoding = "utf-8"?>
<LinearLayout xmlns:android = "http://schemas.android.com/apk/res/android"
    android:layout_width = "wrap_content"
    android:layout_height = "wrap_content"
    android:background = "@drawable/background02"
    android:orientation = "horizontal" >

    <TextView
        android:layout_width = "wrap_content"
        android:layout_height = "wrap_content"
        android:height = "50px"
        android:text = "成绩: " />

    <RadioGroup
        android:id = "@ + id/radioGroup1"
        android:layout_width = "wrap_content"
```

```
        android:layout_height = "wrap_content"
        android:orientation = "horizontal" >

        <RadioButton
            android:id = "@ + id/radio0"
            android:layout_width = "wrap_content"
            android:layout_height = "wrap_content"
            android:checked = "true"
            android:text = "合格" />

        <RadioButton
            android:id = "@ + id/radio1"
            android:layout_width = "wrap_content"
            android:layout_height = "wrap_content"
            android:text = "不合格" />
    </RadioGroup>

    <Button
        android:id = "@ + id/button1"
        android:layout_width = "wrap_content"
        android:layout_height = "wrap_content"
        android:text = "提交" >
    </Button>

</LinearLayout>
```

步骤 6：为"提交"按钮设置响应事件。在 MainActivity 文件中使用 onCheckedChange()方法获取选中单选按钮的值，具体代码如下：

```
final RadioGroup sex = (RadioGroup) findViewById(R.id.radioGroup1);    //获取单选按钮组
//为单选按钮组添加事件监听
sex.setOnCheckedChangeListener(new OnCheckedChangeListener() {
            @Override
            public void onCheckedChanged(RadioGroup group, int checkedId) {
                RadioButton r = (RadioButton) findViewById(checkedId);
Log.i("单选按钮", "选择为: " + r.getText());
            }
        });
```

步骤 7：为"提交"按钮设置事件监听器。在 MainActivity 文件中重写 onClick()方法，并通过遍历的方式获取被选中的按钮值，获取情况在日志面板中可以查到，如图 6-10 所示。其具体代码如下：

```
Button button = (Button) findViewById(R.id.button1);
            button.setOnClickListener(new OnClickListener() {
            @Override
            public void onClick(View v) {
                            for (int i = 0; i < sex.getChildCount(); i++) {
                        RadioButton r = (RadioButton) sex.getChildAt(i);
                        if (r.isChecked()) {/
                        Log.i("单选按钮", "性别:" + r.getText());
                        break;
                    }
                }
            }
        });
```

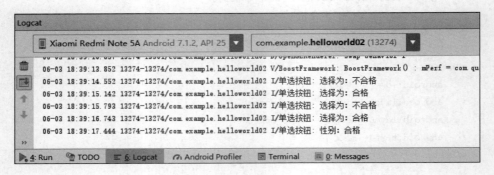

图 6-10　日志面板中选中单选按钮组时获取的值

6.3.5　在 UI 中设计显示图片：ImageView 组件实例

【**例 6.5**】　在 Android Studio 中创建名为 6.5 的 App 项目，使用 ImageView 组件显示相关图像的不同比例效果，包括原始尺寸、限制最大高度和宽度、保持纵横像素比显示全图。

视频讲解

步骤 1：在 Android Studio 中创建一个名为 6.5 的 App 项目，并在项目文件中找到 res 下 layout 文件夹中的 main.xml 文件。

步骤 2：对 main.xml 文件进行编写。为了单选按钮的效果在手机界面上看起来更美观，将新建项目中默认的垂直布局改为水平布局的方式。

步骤 3：将 main.xml 文件中的原有 TextView 组件删除，添加 ImageView 组件，并设置其显示原始尺寸图像，具体代码如下：

```
<?xml version = "1.0" encoding = "utf-8"?>
< LinearLayout xmlns:android = "http://schemas.android.com/apk/res/android"
    android:orientation = "horizontal"
    android:layout_width = "fill_parent"
    android:layout_height = "fill_parent"
    android:background = "@drawable/background"
    >
    < ImageView
        android:src = "@drawable/img1"
        android:id = "@ + id/imageView1"
        android:layout_margin = "5px"
        android:layout_height = "wrap_content"
        android:layout_width = "wrap_content"/>
</LinearLayout >
```

步骤 4：在水平布局管理器中再添加一个 ImageView 组件，并设置其图像显示的最大高度和宽度，具体代码如下：

```
< ImageView
    android:src = "@drawable/img1"
    android:id = "@ + id/imageView2"
    android:maxWidth = "90px"
    android:maxHeight = "90px"
    android:adjustViewBounds = "true"
    android:layout_margin = "5px"
    android:layout_height = "wrap_content"
    android:layout_width = "wrap_content"/>
```

步骤 5：在水平布局管理器中再添加一个 ImageView 组件，并设置其图像在保持纵横像素比的前提下完全显示该图像，具体代码如下：

```
< ImageView
    android:src = "@drawable/img1"
    android:id = "@ + id/imageView3"
    android:scaleType = "fitEnd"      ──► 保持纵横像素比缩放图片
    android:layout_margin = "5px"
    android:layout_height = "90px"
    android:layout_width = "90px"/>
```

步骤 6：运行程序，效果如图 6-11 所示。

6.3.6 在 UI 中设计列表：ListView 组件实例

【例 6.6】 在 Android Studio 中创建名为 6.6 的 App 项目，使用

视频讲解

图 6-11 使用 ImageView 组件显示图片的不同效果

ListView 组件在布局管理器中添加列表视图。

步骤1：在 Android Studio 中创建一个名为 6.6 的 App 项目，并在项目文件中找到 res 下 layout 文件夹中的 main.xml 文件，添加一个 ListView。其具体代码如下：

```xml
<?xml version = "1.0" encoding = "utf - 8"?>
<LinearLayout xmlns:android = "http://schemas.android.com/apk/res/android"
    android:orientation = "vertical"
    android:layout_width = "fill_parent"
    android:layout_height = "fill_parent"
    >
    <ListView android:id = "@ + id/listView1"
        android:entries = "@array/newlist"    ——→ 新建的ListView的名称为newlist
        android:layout_height = "wrap_content"
        android:layout_width = "match_parent"/>
</LinearLayout>
```

步骤2：创建一个用于定义数组资源的 XML 文件，即在 res 下的 values 文件夹中新建 arrays.xml 文件，并在其中添加名为 newlist 的字符串数组。其具体代码如下：

```xml
<?xml version = "1.0" encoding = "utf-8"?>
<resources>
    <string-array name = "newlist">
        <item>智能体感</item>
        <item>多屏互动</item>
        <item>分屏多任务</item>
        <item>单手操作</item>
        <item>超级截屏</item>
        <item>快捷启动</item>
    </string-array>
</resources>
```

步骤 3：运行程序，效果如图 6-12 所示。

图 6-12　ListView 组件实例运行效果图

6.3.7　在 UI 中设计列表选择框：Spinner 组件实例

视频讲解

【例 6.7】　在 Android Studio 中创建名为 6.7 的 App 项目，使用 Spinner 组件在布局管理器中添加列表选择框。

步骤1：在 Android Studio 中创建一个名为 6.7 的 App 项目，并在项目文件中找到 res 下 layout 文件夹中的 main.xml 文件，添加一个 Spinner。其具体代码如下：

```xml
<?xml version = "1.0" encoding = "utf-8"?>
<LinearLayout xmlns:android = "http://schemas.android.com/apk/res/android"
    android:orientation = "horizontal"
    android:layout_width = "fill_parent"
    android:layout_height = "fill_parent"
    >
    <TextView android:id = "@+id/textView1"
        android:text = "请选择图书类型："
        android:layout_height = "wrap_content"
        android:layout_width = "wrap_content"/>
    <Spinner
        android:entries = "@array/newspinner"    →新建Spinner的名称为newspinner
        android:layout_height = "wrap_content"
        android:layout_width = "wrap_content"
        android:id = "@+id/spinner1"/>
    <Button android:text = "提交"
        android:id = "@+id/button1"
        android:layout_width = "wrap_content"
        android:layout_height = "wrap_content"/>
</LinearLayout>
```

步骤2：创建一个用于定义数组资源的 XML 文件，即在 res 下的 values 文件夹中新建 arrays.xml 文件，并在其中添加名为 newspinner 的字符串数组。其具体代码如下：

```xml
<?xml version = "1.0" encoding = "utf-8"?>
<resources>
    <string-array name = "newspinner">
        <item>人文类学科</item>
        <item>自然类学科</item>
        <item>基础学科</item>
        <item>应用学科</item>
        <item>泛论学科</item>
    </string-array>
</resources>
```

步骤3：运行程序，效果如图 6-13 所示。

6.3.8 在 UI 中设计网格视图：GridView 组件实例

视频讲解

【例 6.8】 在 Android Studio 中创建名为 6.8 的 App 项目，使用 GridView

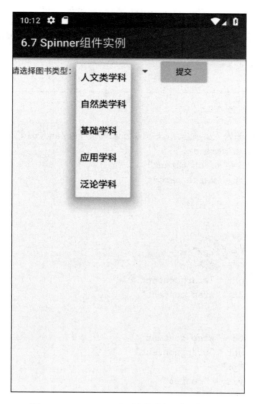

图 6-13　Spinner 组件实例运行效果图

组件在布局管理器中添加图片及相关说明文字的网格视图。

步骤 1：在 Android Studio 中创建一个名为 6.8 的 App 项目，并在项目文件中找到 res 下 layout 文件夹中的 main.xml 文件，添加一个 GridView，然后设置每行显示 4 张图片。其具体代码如下：

```xml
<?xml version = "1.0" encoding = "utf-8"?>
<LinearLayout xmlns:android = "http://schemas.android.com/apk/res/android"
    android:orientation = "vertical"
    android:layout_width = "fill_parent"
    android:layout_height = "fill_parent"
    android:id = "@ + id/llayout"
    >
<GridView android:id = "@ + id/gridView1"
    android:layout_height = "wrap_content"
    android:layout_width = "match_parent"
    android:stretchMode = "columnWidth"
    android:numColumns = "4"></GridView>
</LinearLayout>
```

步骤2：创建一个用于定义GridView的XML文件，即在res下的layout文件夹中新建items.xml文件，并在其中添加一个ImageView和一个TextView组件，实现图片和文字的显示。其具体代码如下：

```xml
<?xml version = "1.0" encoding = "utf-8"?>
<LinearLayout
    xmlns:android = "http://schemas.android.com/apk/res/android"
    android:orientation = "vertical"
    android:layout_width = "match_parent"
    android:layout_height = "match_parent"
    >

<ImageView
    android:id = "@+id/image"
    android:paddingLeft = "12px"
    android:scaleType = "fitCenter"
    android:layout_height = "wrap_content"
    android:layout_width = "wrap_content"
    />
<TextView
    android:layout_width = "wrap_content"
    android:layout_height = "wrap_content"
    android:padding = "6px"
    android:layout_gravity = "center"
    android:id = "@+id/title"
    />
</LinearLayout>
```

步骤3：在MainActivity.java文件中获取图片和文字信息，并与新建的GridView进行关联。其具体代码如下：

```java
package com.mingrisoft;
import java.util.ArrayList;
import java.util.HashMap;
import java.util.List;
import java.util.Map;
import android.app.Activity;
import android.os.Bundle;
import android.widget.GridView;
import android.widget.SimpleAdapter;

public class MainActivity extends Activity {
    private int[] imageId = new int[] { R.drawable.img01, R.drawable.img02,
            R.drawable.img03, R.drawable.img04, R.drawable.img05,
            R.drawable.img06, R.drawable.img07, R.drawable.img08,
            R.drawable.img09, R.drawable.img10, R.drawable.img11,
            R.drawable.img12, };
    @Override
```

```java
public void onCreate(Bundle savedInstanceState) {
    super.onCreate(savedInstanceState);
    setContentView(R.layout.main);
    GridView gridview = (GridView) findViewById(R.id.gridView1);
    String[] title = new String[] { "1号车", "2号车", "3号车", "4号车", "5号车",
            "6号车", "7号车", "8号车", "9号车", "10号车", "11号车", "12号车" };
    List<Map<String, Object>> listItems = new ArrayList<Map<String, Object>>();
    for (int i = 0; i < imageId.length; i++) {
        Map<String, Object> map = new HashMap<String, Object>();
        map.put("image", imageId[i]);
        map.put("title", title[i]);
        listItems.add(map);
    }

    SimpleAdapter adapter = new SimpleAdapter(this,
                    listItems,
                    R.layout.items,
                    new String[] { "title", "image" },
                    new int[] {R.id.title, R.id.image }
    );

    gridview.setAdapter(adapter);
```

步骤4：运行程序，效果如图6-14所示。

图6-14　GridView组件实例运行效果图

6.3.9 在 UI 中设计进度条：ProgressBar 组件实例

【例 6.9】 在 Android Studio 中创建名为 6.9 的 App 项目,使用 ProgressBar 组件在布局管理器中添加水平进度条和圆形进度条。

步骤 1：在 Android Studio 中创建一个名为 6.9 的 App 项目,并在项目文件中找到 res 下 layout 文件夹中的 activity_main.xml 文件,删除默认自带的 TextView 组件,添加一个水平进度条和一个圆形进度条。其具体代码如下：

视频讲解

```xml
<?xml version = "1.0" encoding = "utf-8"?>
<LinearLayout xmlns:android = "http://schemas.android.com/apk/res/android"
    android:orientation = "vertical"
    android:layout_width = "fill_parent"
    android:layout_height = "fill_parent"
    >
<ProgressBar
    android:id = "@ + id/progressBar1"
    android:layout_width = "match_parent"
    android:max = "100"
    style = "@android:style/Widget.ProgressBar.Horizontal"      //创建水平进度条
    android:layout_height = "wrap_content"/>
<ProgressBar
    android:id = "@ + id/progressBar2"
    style = "?android:attr/progressBarStyleLarge"                //创建圆形进度条
    android:layout_width = "wrap_content"
    android:layout_height = "wrap_content"/>
</LinearLayout>
```

步骤 2：在 MainActivity.java 文件中创建两个 ProgressBar 类的对象,分别为水平进度条和圆形进度条。其具体代码如下：

```java
package com.mingrisoft;
import android.app.Activity;
import android.os.Bundle;
import android.os.Handler;
import android.os.Message;
import android.view.View;
import android.widget.ImageView;
import android.widget.ProgressBar;
import android.widget.Toast;
```

```java
public class MainActivity extends Activity {
    private ProgressBar horizonP;           // 创建水平进度条
    private ProgressBar circleP;            // 创建圆形进度条
    private int mProgressStatus = 0;
    private Handler mHandler;
    @Override
    public void onCreate(Bundle savedInstanceState) {
        super.onCreate(savedInstanceState);
        setContentView(R.layout.main);
        horizonP = (ProgressBar) findViewById(R.id.progressBar1);
        circleP = (ProgressBar)findViewById(R.id.progressBar2);
        circleP.incrementProgressBy(-10);
        circleP.setVisibility(View.VISIBLE);
        mHandler = new Handler(){
            @Override
            public void handleMessage(Message msg) {
                if(msg.what == 0x111){
                    horizonP.setProgress(mProgressStatus);
                    circleP.setProgress(100 - mProgressStatus);
                }else{
                    Toast.makeText(MainActivity.this, "操作已经完成",
Toast.LENGTH_SHORT).show();
                    horizonP.setVisibility(View.GONE);
                    circleP.setVisibility(View.GONE);}
            }

        };
        new Thread(new Runnable() {
            public void run() {
                while (true) {
                    mProgressStatus = doWork();
                    Message m = new Message();
                    if(mProgressStatus < 100){
                        m.what = 0x111;
                        mHandler.sendMessage(m);
                    }else{
                        m.what = 0x110;
                        mHandler.sendMessage(m);
                        break;
                    }
                }
            }
            private int doWork() {
                mProgressStatus += Math.random() * 10;
                try {
                    Thread.sleep(200);
                } catch (InterruptedException e) {
                    e.printStackTrace();
```

```
                }
                return mProgressStatus;
            }
        }).start();
    }
}
```

步骤3：运行程序,效果如图 6-15 所示。

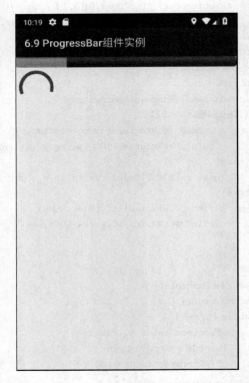

图 6-15　ProgressBar 组件实例运行效果图

6.3.10　在 UI 中设计滑动条：SeekBar 组件实例

【例 6.10】 在 Android Studio 中创建名为 6.10 的 App 项目,使用 SeekBar 组件在布局管理器中添加滑动条。

步骤1：在 Android Studio 中创建一个名为 6.10 的 App 项目,并在项目文件中找到 res 下 layout 文件夹中的 activity_main.xml 文件,在 TextView 组件中修改 text 属性的值,然后通过<SeekBar></SeekBar>标记添加一个滑动条。其具体代码如下：

视频讲解

```xml
<?xml version="1.0" encoding="utf-8"?>
<LinearLayout xmlns:android="http://schemas.android.com/apk/res/android"
    android:orientation="vertical"
    android:layout_width="fill_parent"
    android:layout_height="fill_parent"
    >
    <TextView
        android:text="当前值：60"
        android:id="@+id/textView1"
        android:layout_width="wrap_content"
        android:layout_height="wrap_content"/>
    <SeekBar
        android:layout_height="wrap_content"
        android:id="@+id/seekBar1"
        android:max="100"
        android:progress="50"
        android:padding="10px"
        android:layout_width="match_parent"/>    ───► 添加一个SeekBar
</LinearLayout>
```

步骤2：在MainActivity.java文件中定义一个SeekBar类的对象。其具体代码如下：

```java
package com.mingrisoft;
import android.app.Activity;
import android.os.Bundle;
import android.widget.SeekBar;
import android.widget.Toast;
import android.widget.SeekBar.OnSeekBarChangeListener;
import android.widget.TextView;

public class MainActivity extends Activity {
    private SeekBar seekbar;
    @Override
    public void onCreate(Bundle savedInstanceState) {
        super.onCreate(savedInstanceState);
        setContentView(R.layout.main);
        final TextView result = (TextView)findViewById(R.id.textView1);
        seekbar = (SeekBar) findViewById(R.id.seekBar1);seekbar.setOnSeekBarChangeListener(new OnSeekBarChangeListener() {
            @Override
            public void onStopTrackingTouch(SeekBar seekBar) {
                Toast.makeText(MainActivity.this, "结束滑动", Toast.LENGTH_SHORT).show();
            }

            @Override
            public void onStartTrackingTouch(SeekBar seekBar) {
```

```
                    Toast.makeText(MainActivity.this, "开始滑动",
Toast.LENGTH_SHORT).show();
                }

                @Override
                public void onProgressChanged(SeekBar seekBar, int progress,
                    boolean fromUser) {
                    result.setText("当前值: " + progress);
                }
            });
        }
    }
```

步骤3：运行程序，效果如图6-16所示。

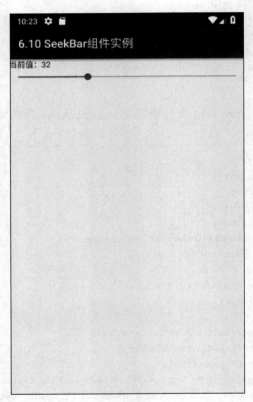

图 6-16 SeekBar 组件实例运行效果图

6.3.11 在 UI 中设计星级评价条：RatingBar 组件实例

视频讲解

【例6.11】 在 Android Studio 中创建名为 6.11 的 App 项目，使用 RatingBar 组件在布局管理器中添加星级评价条。

步骤 1：在 Android Studio 中创建一个名为 6.11 的 App 项目，并在项目文件中找到 res 下 layout 文件夹中的 activity_main.xml 文件，将原有默认的 TextView 组件删除，然后通过< RatingBar ></RatingBar >标记添加一个星级评价条。其具体代码如下：

```xml
<?xml version = "1.0" encoding = "utf-8"?>
< LinearLayout xmlns:android = "http://schemas.android.com/apk/res/android"
    android:orientation = "vertical"
    android:layout_width = "fill_parent"
    android:layout_height = "fill_parent"
    >
    < RatingBar
        android:id = "@ + id/ratingBar1"
        android:numStars = "6"
        android:rating = "3.5"
        android:isIndicator = "false"
        android:layout_width = "wrap_content"
        android:layout_height = "wrap_content"/>
    < Button
        android:text = "提交"
        android:id = "@ + id/button1"
        android:layout_width = "wrap_content"
        android:layout_height = "wrap_content"/>
</LinearLayout >
```

（添加一个"提交"按钮）

步骤 2：在 MainActivity.java 文件中定义一个 RatingBar 类的对象。其具体代码如下：

```java
package com.mingrisoft;

import android.app.Activity;
import android.os.Bundle;
import android.util.Log;
import android.view.View;
import android.view.View.OnClickListener;
import android.widget.Button;
import android.widget.RatingBar;
import android.widget.Toast;

public class MainActivity extends Activity {
    private RatingBar ratingbar;    //定义星级评价条

    @Override
    public void onCreate(Bundle savedInstanceState) {
        super.onCreate(savedInstanceState);
        setContentView(R.layout.main);
```

```
            ratingbar = (RatingBar) findViewById(R.id.ratingBar1);
            Button button = (Button)findViewById(R.id.button1);
button.setOnClickListener(new OnClickListener() {

                @Override
                public void onClick(View v) {
                    int result = ratingbar.getProgress();        ──▶ 获取每次最少改变星
                    float rating = ratingbar.getRating();         ──▶ 获取星级评价条进度
                    float step = ratingbar.getStepSize();         ──▶ 获取星级评价等级进度数量
                    Log.i("星级评价条","step = " + step + " result = " + result + " rating = " + rating);
                    Toast.makeText(MainActivity.this, "斩获" + rating + "颗星", Toast.LENGTH_SHORT).show();

                }
            });

        }
    }
```

步骤 3：运行程序,效果如图 6-17 所示。

图 6-17 RatingBar 组件实例运行效果图

6.3.12 在 UI 中设计图片查看器：ImageSwitcher 组件实例

【例 6.12】 在 Android Studio 中创建名为 6.12 的 App 项目，使用 ImageSwitcher 组件在布局管理器中实现图片的查看功能。

步骤 1：在 Android Studio 中创建一个名为 6.12 的 App 项目，并在项目文件中找到 res 下 layout 文件夹中的 activity_main.xml 文件，将原有默认的 TextView 组件删除，然后通过<ImageSwitcher></ImageSwitcher>标记添加一个图像切换器，并添加两个按钮。其具体代码如下：

视频讲解

```
<?xml version = "1.0" encoding = "utf - 8"?>
<LinearLayout xmlns:android = "http://schemas.android.com/apk/res/android"
    android:orientation = "horizontal"
    android:layout_width = "fill_parent"
    android:layout_height = "fill_parent"
    android:id = "@ + id/llayout"
    android:gravity = "center"
    >
    <Button
        android:text = "BACK"
        android:id = "@ + id/button1"
        android:layout_width = "wrap_content"
        android:layout_height = "wrap_content"/>         ——▶ 添加BACK按钮
    <ImageSwitcher
        android:id = "@ + id/imageSwitcher1"
        android:layout_gravity = "center"
        android:layout_width = "wrap_content"
        android:layout_height = "wrap_content"/>         ——▶ 添加一个ImageSwitcher
    <Button
        android:text = "NEXT"
        android:id = "@ + id/button2"
        android:layout_width = "wrap_content"
        android:layout_height = "wrap_content"/>         ——▶ 添加NEXT按钮
</LinearLayout>
```

步骤 2：在 MainActivity.java 文件中定义一个用于保存图像 id 的数组，再定义一个 ImageSwitcher 类的对象。其具体代码如下：

```
package com.mingrisoft;
```

```java
import android.app.Activity;
import android.os.Bundle;
import android.view.View;
import android.view.View.OnClickListener;
import android.view.ViewGroup.LayoutParams;
import android.view.animation.AnimationUtils;
import android.widget.Button;
import android.widget.ImageSwitcher;
import android.widget.ImageView;
import android.widget.ViewSwitcher.ViewFactory;

public class MainActivity extends Activity {
    private int[] imageId = new int[] { R.drawable.img01, R.drawable.img02,
            R.drawable.img03, R.drawable.img04, R.drawable.img05,
            R.drawable.img06, R.drawable.img07, R.drawable.img08,
            R.drawable.img09 };
    private int index = 0;
    private ImageSwitcher imageSwitcher;

    @Override
    public void onCreate(Bundle savedInstanceState) {
        super.onCreate(savedInstanceState);
        setContentView(R.layout.main);
        imageSwitcher = (ImageSwitcher) findViewById(R.id.imageSwitcher1);
        imageSwitcher.setInAnimation(AnimationUtils.loadAnimation(this,
                android.R.anim.fade_in));
        imageSwitcher.setOutAnimation(AnimationUtils.loadAnimation(this,
                android.R.anim.fade_out));
        imageSwitcher.setFactory(new ViewFactory() {

            @Override
            public View makeView() {
                ImageView imageView = new ImageView(MainActivity.this);
                imageView.setScaleType(ImageView.ScaleType.FIT_CENTER);
                imageView.setLayoutParams(new ImageSwitcher.LayoutParams(
LayoutParams.WRAP_CONTENT, LayoutParams.WRAP_CONTENT));
                return imageView;
            }

        });
        imageSwitcher.setImageResource(imageId[index]);
        Button up = (Button) findViewById(R.id.button1);
        Button down = (Button) findViewById(R.id.button2);
        up.setOnClickListener(new OnClickListener() {

            @Override
            public void onClick(View v) {

                if (index > 0) {
```

声明一个保存图像的数组

```
                    index--;
            } else {
                index = imageId.length - 1;
            }
            imageSwitcher.setImageResource(imageId[index]);
        }
    });
    down.setOnClickListener(new OnClickListener() {

        @Override
        public void onClick(View v) {
            if (index < imageId.length - 1) {
                index++;
            } else {
                index = 0;
            }
            imageSwitcher.setImageResource(imageId[index]);
        }
    });
}
```

步骤 3：运行程序，效果如图 6-18 所示。

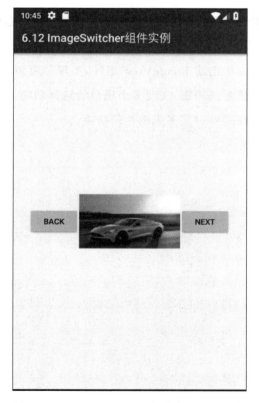

图 6-18　ImageSwitcher 组件实例运行效果图

6.4 项目结案

本项目通过实例的方式为大家介绍了 Android 常用的 UI 组件,包括文本框、编辑框、按钮、图像视图、单选按钮、日期、计时器等基本组件,以及文本框、拖动条、列表选择框、网格视图、进度条、星级评价条、图像切换器等高级组件。通过这些组件的创建与运用能够很好地进行 App 的 UI 设计,希望大家能够举一反三,将 App 的 UI 界面设计从技术层面逐渐向艺术层面转化,真正实现活学活用。

6.5 项目练习

1. 开发一款 App,在其中创建 TextView 组件,实现在手机界面上添加文字说明的效果。
2. 开发一款 App,在其中创建 ImageView 组件,实现带边框的图片的显示。
3. 开发一款 App,设置复选按钮,实现多个项目的选择功能。
4. 开发一款 App,实现能够预览多张图片的功能。

项目7

设置App的多媒体应用

7.1 项目目标：为App添加多媒体应用

对于一个完善的App而言,多媒体的适当应用能够使App的功能更加丰富,与用户之间的沟通更加亲密。音频和视频在传达信息方面比单纯的文字与图片更具吸引力和突出性。那么如何在App界面上添加多媒体,如何控制多媒体的运行,即为本项目的主要内容。本项目通过结合实例讲解,希望能够使大家轻松地使用多媒体,为App的设计添加交互强的多媒体控制功能。

7.2 项目准备

7.2.1 介绍音频控制类

对于音频的控制,在Android开发中提供了MediaPlayer类,能够方便开发者对音频进行播放、暂停、继续和停止等功能的实现。其实现方法十分简单,只要在创建该类对象的基础上进行相关方法的调用即可。

1. 创建对象

MediaPlayer 类仅有一个无参的构造函数,为了方便对象的初始化,提供了几个静态的 create()方法用于完成 MediaPlayer 初始化的工作。其常用的方法有以下两种。

(1) static MediaPlayer create(Context context, int resid):通过资源的 id(即音频资源的绝对路径)来创建 MediaPlayer 对象。

(2) static MediaPlayer create(Context context, Uri uri):通过指定 Uri(即音频资源的网络地址)来创建 MediaPlayer 对象。

与此同时,创建 MediaPlayer 对象也可以采用 new 的方式,例如:

```
MediaPlayer mp = new MediaPlayer();
```

2. 音频文件的装载

在播放音频文件之前,首先要完成音频的装载。装载音频的方法有两种,分别是 MediaPlayer 类和 Sound Pool 类,更加适用与合理的方法为 prepareAsync()的异步装载方式。其具体代码如下:

```
MediaPlayer mp = new MediaPlayer();
mp.setDataSource(path);
mp.setAudioStreamType(AudioManager.STREAM_MUSIC);        //通过异步的方式装载媒体资源
mp.prepareAsync();          ——→ 异步装载音频 mp
mp.setOnPreparedListener(new OnPreparedListener() {
    @Override
    public void onPrepared(MediaPlayer mp) {
        mp.start();         ——→ 装载后即可播放音频 mp
    }
});
```

3. 播放、暂停、停止音频

对于音频的控制,可以使用 void pause()暂停、使用 void start()开始、使用 void stop() 停止。以上面创建名为 mp 的 MediaPlayer 类为例,使用方法的具体代码如下。

(1) "mp.pause();":对象 mp 暂停播放。

(2) "mp.start();":对象 mp 开始播放。

(3)"mp.stop();":对象 mp 停止播放。

7.2.2 介绍视频控制类

1. VideoView 组件

对于视频的控制,Android 中自带的 VideoView 组件可以实现各项控制功能。在使用的过程中,首先要创建 VideoView 组件的对象,然后通过 Activity 获取该对象,在加载视频对象后,再通过与控制音频相同的方法来控制视频,例如播放视频使用 start()方法,暂停视频使用 pause()方法,停止视频使用 stop()方法等,常见方法如表 7-1 所示。

表 7-1 VideoView 组件调用方法及使用说明

方法名称	使用说明
setVideoPath()	设置播放视频的位置
start()	开始或继续播放视频
pause()	暂停播放视频
resume()	将视频从头开始播放
seekTo()	从指定的位置开始播放视频
isPlaying()	判断当前是否正在播放视频
getDuration()	获取载入的视频文件的时长

在界面上添加 VideoView 组件,可以通过<VideoView></VideoView>标记来编写。其具体代码如下:

```
<VideoView
    属性列表
    >
</VideoView>
```

VideoView 组件支持的 XML 属性如表 7-2 所示。

表 7-2 VideoView 组件支持的 XML 属性

XML 属性	使用说明
android:id	设置视频组件的 id
android:layout_width	设置视频的宽度
android:layout_heigh	设置视频的高度
android:background	设置视频的播放背景
android:layout_gravity	设置视频的对齐方式

2. SurfaceView 组件

在开发 Android 时,除了使用 VideoView 组件实现对视频的控制以外,还可以通过 SurfaceView 组件配合 MediaPlayer 组件实现对视频的控制。具体的实现步骤如下:

第 1 步:在 main.xml 文件中定义 SurfaceView 组件。

第 2 步:在 MainActivity.java 文件中声明 MediaPlayer 对象和 SurfaceView 对象。

第 3 步:实例化 MediaPlayer 对象,并添加 SurfaceView 组件。

第 4 步:使用 MediaPlayer 组件配套的方法实现对视频的控制,例如播放等。

7.2.3 介绍相机控制类

对于手机而言,相机的使用是必不可少的。在 App 中如果能够结合相机的使用,则能够为其功能增色许多。在 Android 开发中,Camera 类专门用于控制相机以实现相机拍照和录像功能,其中包含很多控制相机的方法,如表 7-3 所示。

表 7-3 Camera 类常用方法及说明

方 法	说 明
Camera open()	打开 Camera,返回一个 Camera 实例
Camera open(int cameraId)	根据 cameraId 打开一个 Camera,返回一个 Camera 实例
release()	释放 Camera 的资源
getNumberOfCameras()	获取当前设备支持的 Camera 硬件个数
Camera.Parameters getParameters()	获取 Camera 的各项参数设置类
setParameters(Camera.Parameters params)	通过 params 把 Camera 的各项参数写入 Camera 中
setDisplayOrientation(int degrees)	摄像预览的旋转度
setPreviewDisplay(SurfaceHolder holder)	设置 Camera 预览的 SurfaceHolder
startPreview()	开始 Camera 的预览
stopPreview()	停止 Camera 的预览
autoFocus(Camera.AutoFocusCallback cb)	自动对焦
takePicture(Camera.ShutterCallback shutter, Camera.PictureCallback raw, Camera.PictureCallback jpeg)	拍照
lock()	锁定 Camera 硬件,使其他应用无法访问
unlock()	解锁 Camera 硬件,使其他应用可以访问

在开发 App 使用相机之前,首先需要在 Manifest 文件中申请权限。根据诉求的不同,申请的方式也有所不同,具体方法如表 7-4 所示。

表7-4 Camera类申请权限及相关方法

申 请 名 称	申 请 方 法	说　　明
Camera Permission	< uses-permission android:name = "android.permission.CAMERA" />	申请使用设备相机
Storage Permission	< uses-permission android:name = "android.permission.WRITE_EXTERNAL_STORAGE" />	申请照片或者视频存储到设备中
Audio Recording Permission	< uses-permission android:name = "android.permission.RECORD_AUDIO" />	申请录音权限
Location Permission	< uses-permission android:name = "android.permission.ACCESS_FINE_LOCATION" />	申请记录地理位置

7.3 项目运行

7.3.1 设计音频控制

【例7.1】 在Android Studio中创建名为7.1的App项目,使用MediaPlayer类在布局管理器中实现对音频的控制,包括播放、暂停、继续和停止等。

步骤1:将需要播放的音频资源放置到手机的SD卡的根目录中,音频文件的名字为000.mp3。

步骤2:在Android Studio中创建一个名为7.1的App项目,并在项目文件中找到res下layout文件夹中的activity_main.xml文件,添加控制音频的Play、Pause、Stop按钮。其具体代码如下:

视频讲解

```xml
<?xml version = "1.0" encoding = "utf-8"?>
< LinearLayout xmlns:android = "http://schemas.android.com/apk/res/android"
    android:layout_width = "fill_parent"
    android:layout_height = "fill_parent"
    android:orientation = "vertical" >

    < TextView
        android:id = "@ + id/hint"
        android:layout_width = "wrap_content"
        android:layout_height = "wrap_content"
```

```xml
        android:padding = "10px"
        android:text = "从单击Play开始播放音频" />

    <LinearLayout
        android:id = "@+id/linearLayout1"
        android:layout_width = "match_parent"
        android:layout_height = "wrap_content" >
        <Button
            android:id = "@+id/button1"
            android:layout_width = "wrap_content"
            android:layout_height = "wrap_content"
            android:text = "Play" />               ← 添加Play按钮

        <Button
            android:id = "@+id/button2"
            android:layout_width = "wrap_content"
            android:layout_height = "wrap_content"
            android:enabled = "false"
            android:text = "Pause" />              ← 添加Pause按钮

        <Button
            android:id = "@+id/button3"
            android:layout_width = "wrap_content"
            android:layout_height = "wrap_content"
            android:enabled = "false"
            android:text = "Stop" />               ← 添加Stop按钮
    </LinearLayout>
</LinearLayout>
```

步骤3：在MainActivity.java文件中定义一个MediaPlayer类的对象。其具体代码如下：

```java
package com.example.helloworld02;

import java.io.File;

import android.app.Activity;
import android.database.Cursor;
import android.media.MediaPlayer;
import android.media.MediaPlayer.OnCompletionListener;
import android.net.Uri;
import android.os.Bundle;
import android.provider.MediaStore;
import android.support.v7.app.AppCompatActivity;
import android.view.View;
import android.view.View.OnClickListener;
import android.widget.Button;
import android.widget.TextView;
```

```java
public class MainActivity extends AppCompatActivity {
    private MediaPlayer mp;
    private boolean isPause = false;
    private String musicPath;
    private TextView hint;
    @Override
    public void onCreate(Bundle savedInstanceState) {
        super.onCreate(savedInstanceState);
        setContentView(R.layout.activity_main);
        final Button button1 = (Button) findViewById(R.id.button1);
        final Button button2 = (Button) findViewById(R.id.button2);
        final Button button3 = (Button) findViewById(R.id.button3);
        hint = (TextView) findViewById(R.id.hint);
        file = new File ("/sdcard/000.mp3");                    // 音频地址
        String[] strings = {
                MediaStore.Audio.Media.TITLE,
                MediaStore.Audio.Media.ARTIST,
                MediaStore.Audio.Media.DATA
        };
        Cursor cursor = this.getContentResolver().query(uri,strings,null,null,
                null);
        cursor.moveToNext();
        musicPath = cursor.getString(2);
        mp = new MediaPlayer();                                  // 定义一个MediaPlayer对象,名为mp
        mp.setOnCompletionListener(new OnCompletionListener() {

            @Override
            public void onCompletion(MediaPlayer mp) {           // 重新播放音频
                play();
            }
        });

        button1.setOnClickListener(new OnClickListener() {

            @Override
            public void onClick(View v) {
                play();
                if (isPause) {
                    button2.setText("Pause");
                    isPause = false;
                }
                button2.setEnabled(true);
                button3.setEnabled(true);
                button1.setEnabled(false);
            }
        });
```

```java
button2.setOnClickListener(new OnClickListener() {
    @Override
    public void onClick(View v) {
        if (mp.isPlaying() && !isPause) {
            mp.pause();
            isPause = true;
            button2.setText("Go on");
            hint.setText("Pause the music");
            button1.setEnabled(true);
        } else {
            mp.start();
            button2.setText("Pause");
            hint.setText("Go on to play the music");
            isPause = false;
            button1.setEnabled(false);
        }
    }
});
```
⟶ 暂停播放音频

```java
button3.setOnClickListener(new OnClickListener() {

    @Override
    public void onClick(View v) {
        mp.stop();
        hint.setText("Stop the music");
        button1.setEnabled(true);
        button2.setEnabled(false);
        button3.setEnabled(false);

    }
});
}
```
⟶ 停止播放音频

```java
private void play() {
    try {
        mp.reset();
        mp.setDataSource(musicPath);
        mp.prepare();
        mp.start();
        hint.setText("The music is playing");
    } catch (Exception e) {
        e.printStackTrace();
    }
}
```
⟶ 播放音频

```
@Override
protected void onDestroy() {
    if (mp.isPlaying()) {
        player.stop();
    }
    mp.release();
    super.onDestroy();
}
```
───→ 释放音频资源

}

步骤4：运行程序，效果如图7-1～图7-3所示。

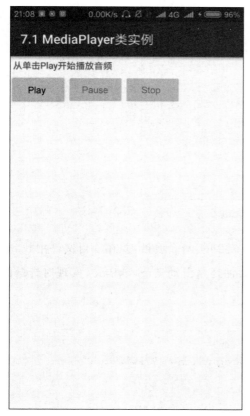

图7-1 音频控制的默认界面

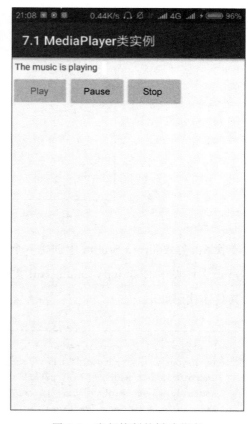

图7-2 音频控制的播放状态

【例7.2】 在Android Studio中创建名为7.2的App项目，并在例7.1的基础上使用MediaPlayer类在布局管理器中实现对音频音量的控制。

步骤1：将需要播放的音频资源放置到手机SD卡的根目录中，音频文件的名字为000.mp3。

视频讲解

图 7-3　音频控制的暂停状态

步骤 2：在 Android Studio 中创建一个名为 7.2 的 App 项目，并在项目文件中找到 res 下 layout 文件夹中的 activity_main.xml 文件，添加音量滑动条（SeekBar），实现对音量的控制。其具体代码如下：

```xml
<?xml version = "1.0" encoding = "utf-8"?>
<LinearLayout xmlns:android = "http://schemas.android.com/apk/res/android"
    android:layout_width = "fill_parent"
    android:layout_height = "fill_parent"
    android:orientation = "vertical" >

    <TextView
        android:id = "@+id/hint"
        android:layout_width = "wrap_content"
        android:layout_height = "wrap_content"
        android:padding = "10px"
        android:text = "单击 Play 播放音频" />

    <LinearLayout
```

```xml
            android:id = "@ + id/linearLayout1"
            android:layout_width = "match_parent"
            android:layout_height = "wrap_content"
            android:orientation = "horizontal">

        < Button
            android:id = "@ + id/button1"
            android:layout_width = "wrap_content"
            android:layout_height = "wrap_content"
            android:text = "Play" />

        < Button
            android:id = "@ + id/button2"
            android:layout_width = "wrap_content"
            android:layout_height = "wrap_content"
            android:enabled = "false"
            android:text = "Pause" />

        < Button
            android:id = "@ + id/button3"
            android:layout_width = "wrap_content"
            android:layout_height = "wrap_content"
            android:enabled = "false"
            android:text = "Stop" />

        < TextView
            android:id = "@ + id/volume"
            android:layout_width = "wrap_content"
            android:layout_height = "wrap_content"
            android:padding = "10px"
            android:text = "Volume: " />

    </LinearLayout >
    < SeekBar
        android:layout_gravity = "top"
        android:id = "@ + id/seekBar1"
        android:layout_width = "match_parent"
        android:layout_height = "wrap_content"
        />
</LinearLayout >
```
→ 添加控制音量滑动条

步骤 3：在 MainActivity.java 文件中设置滑动条音量控制的方法。其关键代码如下：

```
final AudioManager mp = (AudioManager) MainActivity.this.getSystemService(Context.AUDIO_SERVICE);
        MainActivity.this.setVolumeControlStream(AudioManager.STREAM_MUSIC);
        SeekBar seekbar = (SeekBar) findViewById(R.id.seekBar1);
```
→ 设置SeekBar

```
            seekbar.setMax(am.getStreamMaxVolume(AudioManager.STREAM_MUSIC));
            int progress = mp.getStreamVolume(AudioManager.STREAM_MUSIC);
            seekbar.setProgress(progress);
final TextView controlor = (TextView)findViewById(R.id.volume);
     controlor.setText("Volume" + progress);
        seekbar.setOnSeekBarChangeListener(new OnSeekBarChangeListener() {
            @Override
            public void onStopTrackingTouch(SeekBar seekBar) {}
            @Override
            public void onStartTrackingTouch(SeekBar seekBar) {}
            @Override
            public void onProgressChanged(SeekBar seekBar, int progress,boolean fromUser) {
                controlor.setText("Volume" + progress);
        mp.setStreamVolume(AudioManager.STREAM_MUSIC, progress,
            AudioManager.FLAG_PLAY_SOUND);
            }
        });
```

步骤4：运行程序，效果如图7-4所示。

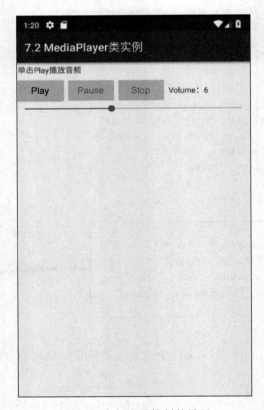

图7-4　音频音量控制的界面

7.3.2 设计视频控制

【例7.3】 在Android Studio中创建名为7.3的App项目,使用VideoView组件在手机上实现对视频的播放控制。

步骤1:在Android Studio中创建一个名为7.3的App项目,并在项目文件中找到res下layout文件夹中的activity_main.xml文件,添加一个VideoView组件,实现对视频的控制。其具体代码如下:

视频讲解

```xml
<?xml version = "1.0" encoding = "utf-8"?>
<LinearLayout xmlns:android = "http://schemas.android.com/apk/res/android"
    android:layout_width = "match_parent"
    android:layout_height = "match_parent"
    android:background = "@drawable/mpbackground"
    android:orientation = "horizontal" >

    <VideoView
        android:id = "@+id/video"
        android:layout_width = "match_parent"
        android:layout_height = "wrap_content"
        android:layout_gravity = "center" />           ——▶ 添加VideoView组件

</LinearLayout>
```

步骤2:在MainActivity.java文件中声明VideoView组件的对象,并对视频添加控制方法。其关键代码如下:

```java
package com.example.helloworld;
import java.io.File;
import android.app.Activity;
import android.media.MediaPlayer;
import android.media.MediaPlayer.OnCompletionListener;
import android.os.Bundle;
import android.os.Environment;
import android.support.v7.app.AppCompatActivity;
import android.util.Log;
import android.widget.MediaController;
import android.widget.Toast;
import android.widget.VideoView;

public class MainActivity extends AppCompatActivity {
```

```java
private VideoView tv;    // 声明VideoView对象
@Override
public void onCreate(Bundle savedInstanceState) {
    super.onCreate(savedInstanceState);
    setContentView(R.layout.activity_main);
    tv = (VideoView) findViewById(R.id.video);
    File file = new File("/storage/4AAC-101B/sdcard/成龙.mp4");    // 获取视频文件
    MediaController mc = new MediaController(MainActivity.this);
    if(file.exists()){
        tv.setVideoPath(file.getAbsolutePath());
        tv.setMediaController(mc);
        tv.requestFocus();
        try {tv.start();
        } catch (Exception e) {
            e.printStackTrace();    // 播放视频文件
        }
        tv.setOnCompletionListener(new OnCompletionListener() {

            @Override
            public void onCompletion(MediaPlayer mp) {
        Toast.makeText(MainActivity.this,"The video is over.",
                        Toast.LENGTH_SHORT).show();
            }
        });
    }else{
        Toast.makeText(this, "The video is not exist. ",
                        Toast.LENGTH_SHORT).show();
            }
    }

}
```

步骤3：运行程序，效果如图7-5所示。

【例7.4】 在 Android Studio 中创建名为 7.4 的 App 项目，使用 SurfaceView 组件配合 MediaPlayer 在手机上实现对视频的播放控制。

步骤1：在 Android Studio 中创建一个名为 7.4 的 App 项目，并在项目文件中找到 res 下 layout 文件夹中的 activity_main.xml 文件，添加一个 SurfaceView 组件和控制视频的按钮。其具体代码如下：

视频讲解

```xml
<?xml version = "1.0" encoding = "utf-8"?>
< LinearLayout xmlns:android = "http://schemas.android.com/apk/res/android"
```

项目7　设置App的多媒体应用

图 7-5　VideoView 组件播放视频实例运行界面

```
android:layout_width = "fill_parent"
android:layout_height = "fill_parent"
android:gravity = "center"
android:orientation = "vertical" >
< SurfaceView
    android:id = "@ + id/surfaceView1"
    android:layout_width = "300dp"
    android:layout_height = "200dp"
    android:keepScreenOn = "true" />
< LinearLayout
    android:id = "@ + id/linearLayout1"
    android:layout_width = "wrap_content"
    android:layout_height = "wrap_content" >
```

——→ 添加SurfaceView组件

```xml
<Button
    android:id = "@+id/play"
    android:layout_width = "wrap_content"
    android:layout_height = "wrap_content"
    android:text = "Play" />                    ——► 添加Play按钮
<Button
    android:id = "@+id/pause"
    android:layout_width = "wrap_content"
    android:layout_height = "wrap_content"
    android:enabled = "false"
    android:text = "Pause" />                   ——► 添加Pause按钮
<Button
    android:id = "@+id/stop"
    android:layout_width = "wrap_content"
    android:layout_height = "wrap_content"
    android:text = "Stop" />                    ——► 添加Stop按钮
    </LinearLayout>
</LinearLayout>
```

步骤2：在MainActivity.java文件中声明SurfaceView组件和MediaPlayer组件的对象，并对视频添加控制方法。其关键代码如下：

```java
package com.example.helloworld;

import java.io.IOException;

import android.app.Activity;
import android.media.MediaPlayer;
import android.media.MediaPlayer.OnCompletionListener;
import android.os.Bundle;
import android.support.v7.app.AppCompatActivity;
import android.view.SurfaceView;
import android.view.View;
import android.view.View.OnClickListener;
import android.widget.Button;
import android.widget.Toast;

public class MainActivity extends AppCompatActivity {
    private MediaPlayer mplayer;      ——► 声明MediaPlayer对象，名为mplayer
    private SurfaceView sview;        ——► 声明SurfaceView对象，名为sview
    @Override
    public void onCreate(Bundle savedInstanceState) {
        super.onCreate(savedInstanceState);
        setContentView(R.layout.activity_main);
        mplayer = new MediaPlayer();
```

```java
        sview = (SurfaceView)findViewById(R.id.surfaceView1);
        Button play = (Button)findViewById(R.id.play);
        final Button pause = (Button)findViewById(R.id.pause);
        Button stop = (Button)findViewById(R.id.stop);
    play.setOnClickListener(new OnClickListener() {
            
            @Override
            public void onClick(View v) {
                mplayer.reset();
try {
        mplayer.setDataSource("/storage/4AAC-101B/sdcard/成龙.mp4");
        mplayer.setDisplay(sv.getHolder());
                mplayer.prepare();
                mplayer.start();
             pause.setText("Pause");
                pause.setEnabled(true);/
            } catch (IllegalArgumentException e) {
                e.printStackTrace();
            } catch (SecurityException e) {
                e.printStackTrace();
            } catch (IllegalStateException e) {
                e.printStackTrace();
            } catch (IOException e) {
                e.printStackTrace();
            }
    
            
        }
    });
    stop.setOnClickListener(new OnClickListener() {
        
        @Override
        public void onClick(View v) {
            if(mplayer.isPlaying()){
                mplayer.stop();
         pause.setEnabled(false);
            }
            
        }
    });
     pause.setOnClickListener(new OnClickListener() {
        
        @Override
        public void onClick(View v) {
            if(mplayer.isPlaying()){
                mplayer.pause();
                ((Button)v).setText("Go on");
            }else{
                mplayer.start();
                ((Button)v).setText("Pause");
```

```
                }
            }
        });
         mp.setOnCompletionListener(new OnCompletionListener() {

            @Override
            public void onCompletion(MediaPlayer mp) {
                Toast.makeText(MainActivity.this, "The video is over",
Toast.LENGTH_SHORT).show();
            }
        });

    }
    @Override
    protected void onDestroy() {
        if(mplayer.isPlaying()){
            mplayer.stop();}
        mp.release();
        super.onDestroy();
    }

}
```

步骤3：运行程序，效果如图7-6所示。

图7-6　SurfaceView组件播放视频实例运行界面

7.3.3 设计相机控制

【例 7.5】 在 Android Studio 中创建名为 7.5 的 App 项目，使用 Camera 组件实现手机控制拍照。

视频讲解

步骤 1：在 Android Studio 中创建一个名为 7.5 的 App 项目，并在项目文件中找到 res 下 layout 文件夹中的 activity_main.xml 文件，添加一个 SurfaceView 组件和控制相机运行的按钮。其具体代码如下：

```xml
<?xml version = "1.0" encoding = "utf-8"?>
<LinearLayout xmlns:android = "http://schemas.android.com/apk/res/android"
    android:layout_width = "fill_parent"
    android:layout_height = "fill_parent"
    android:orientation = "horizontal" >

    <LinearLayout
        android:id = "@ + id/linearLayout1"
        android:layout_width = "72dp"
        android:layout_height = "match_parent"
        android:orientation = "vertical" >

        <Button
            android:id = "@ + id/preview"
            android:layout_width = "wrap_content"
            android:layout_height = "wrap_content"
            android:text = "@string/preview" />     ← 添加预览按钮 PREVIEW

        <Button
            android:id = "@ + id/takephoto"
            android:layout_width = "wrap_content"
            android:layout_height = "wrap_content"
            android:text = "@string/takephoto" />   ← 添加拍照按钮 TAKEPHOTO
    </LinearLayout>

        <SurfaceView
            android:id = "@ + id/surfaceView1"
            android:layout_width = "match_parent"
            android:layout_height = "match_parent" />   ← 添加 SurfaceView 组件

</LinearLayout>
```

步骤 2：在 MainActivity.java 文件中声明 Camera 组件的对象，并对相机添加控制方法。其关键代码如下：

```java
public class MainActivity extends AppCompatActivity {
    private Camera cam;                    // 声明Camera对象，名为cam
    private boolean isPreview = false;
    @Override
    public void onCreate(Bundle savedInstanceState) {
        super.onCreate(savedInstanceState);
        requestWindowFeature(Window.FEATURE_NO_TITLE);
        setContentView(R.layout.activity_main);
            if (!android.os.Environment.getExternalStorageState().equals(
                android.os.Environment.MEDIA_MOUNTED)) {
            Toast.makeText(this, "Please install SD card!", Toast.LENGTH_SHORT).show();
        final SurfaceView sview = (SurfaceView) findViewById(R.id.surfaceView1);
        final SurfaceHolder sholer = sview.getHolder();
        sholer.setType(SurfaceHolder.SURFACE_TYPE_PUSH_BUFFERS);

        Button preview = (Button) findViewById(R.id.preview);
        preview.setOnClickListener(new View.OnClickListener() {
            @Override
            public void onClick(View v1) {
                if (!isPreview) {
                    cam = Camera.open();          // 打开相机
                }

                try {
                    cam.setPreviewDisplay(sholer);
                    Camera.Parameters parameters = cam.getParameters();
                    parameters.setPictureSize(576, 320);
                    parameters.setPictureFormat(PixelFormat.JPEG);
                    parameters.setPictureSize(576, 320);
                    cam.setParameters(parameters);
                    cam.startPreview();
                    cam.autoFocus(null);
                } catch (IOException e) {
                    e.printStackTrace();
                }

            }
        });

        Button takePhoto = (Button) findViewById(R.id.takephoto);
        takePhoto.setOnClickListener(new View.OnClickListener() {
            @Override
            public void onClick(View v1) {
                if(cam!= null){
                    cam.takePicture(null, null, jpeg);
                }
            }
        };
    }
```

控制相机的方法详见表7-3

拍照事件的设置

```java
        final PictureCallback jpeg = new PictureCallback() {
            @Override
            public void onPictureTaken(byte[] data, Camera cam) {
                final Bitmap bmap = BitmapFactory.decodeByteArray(data, 0,
                        data.length);
                View saveImage = getLayoutInflater().inflate(R.layout.save, null);
                final EditText ImageName = (EditText) saveImage
                        .findViewById(R.id.phone_name);
                ImageView v2 = (ImageView) saveImage.findViewById(R.id.show);
                V2.setImageBitmap(bm);
                camstopPreview();
                isPreview = false;

                new AlertDialog.Builder(MainActivity.this).setView(saveImage )
                        .setPositiveButton("Save", new DialogInterface.OnClickListener() {
                            @Override
                            public void onClick(DialogInterface dialog, int which) {
                                File file = new File("/sdcard/" + ImageName.getText().toString() + ".jpg");
                                try {
                                    file.createNewFile();
                                    FileOutputStream fOS = new FileOutputStream(file);
                                    bmap.compress(Bitmap.CompressFormat.JPEG, 80, fOS);
                                    fOS.flush();
                                    fOS.close();
                                    isPreview = true;
                                    resetCamera();
                                } catch (IOException e) {
                                    e.printStackTrace();
                                }
                            }

                        }).setNegativeButton("Cancel",
                        new DialogInterface.OnClickListener() {
                            public void onClick(DialogInterface dialog, int which) {
                                isPreview = true;
                                resetCamera();
                            }
                        }).show();
            }
        };
        private void resetCamera(){
            if(isPreview){
                cam.startPreview();
            }
        }
        @Override
```

使用onPictureTaken()方法将拍下的图片保存为位图

将位图保存在指定文件夹中

```
    protected void onPause() {
        if(cam!= null){
            cam.stopPreview();
            cam.release();
        }
        super.onPause();
    }

}
```

步骤3:为了实现保存照片的界面,需要在 res 下的 layout 文件夹中创建一个 save.xml 文件,其中包含输入照片名字与预览照片的 UI 设计。其具体代码如下:

```
<?xml version = "1.0" encoding = "utf - 8"?>
<LinearLayout xmlns:android = "http://schemas.android.com/apk/res/android"
    android:orientation = "vertical"
    android:layout_width = "fill_parent"
    android:layout_height = "fill_parent">
<LinearLayout
    android:orientation = "horizontal"
    android:layout_width = "fill_parent"
    android:layout_height = "wrap_content">
    <TextView
        android:layout_width = "wrap_content"
        android:layout_height = "wrap_content"
        android:layout_marginRight = "8dp"
        android:text = "Photo's name: "
        />          ← TextView组件用于设置保存照片的文字提示

    <EditText
        android:id = "@ + id/phone_name"
        android:layout_width = "fill_parent"
        android:layout_height = "wrap_content"/>   ← EditText组件用于输入照片名字
</LinearLayout>
    <ImageView
        android:id = "@ + id/show"
        android:contentDescription = "preview the photo"
        android:layout_width = "350dp"
        android:layout_height = "260dp"
        android:scaleType = "fitCenter"
        android:layout_marginTop = "10dp"/>   ← ImageView组件用于设置预览照片
</LinearLayout>
```

步骤 4：申请相机的访问权限，在 AndroidManifest.xml 文件中进行设置。其关键代码如下：

```xml
<?xml version = "1.0" encoding = "utf-8"?>
<manifest xmlns:android = "http://schemas.android.com/apk/res/android"
    package = "com.example.helloworld">
    <uses-permission android:name = "android.permission.MOUNT_UNMOUNT_FILESYSTEMS"/>
    <uses-permission android:name = "android.permission.WRITE_EXTERNAL_STORAGE"/>
    <uses-permission android:name = "android.permission.CAMERA"/>
    <uses-feature android:name = "android.hardware.cam"/>
    <uses-feature android:name = "android.hardware.cam.autofocus"/>
```

申请权限及相关方法，详见表7-4

步骤 5：运行程序，效果如图 7-7 和图 7-8 所示。

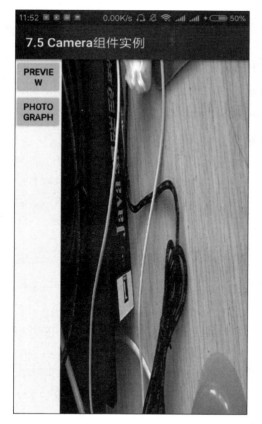

图 7-7　Camera 组件运行相机预览界面

图 7-8　保存照片对话框界面

7.4 项目结案

本项目通过实例的方式为大家介绍了 Android 常用的多媒体控制组件与方法,包括对音频、视频及照相机的使用与调控等。通过实例演示,能够较为清晰地获取如何播放、暂停、停止、装载多媒体文件,也能够在 App 中添加拍照及保存照片等功能,这对于开发 App 而言是十分重要的,也为所设计的 App 增添了趣味性与参与性。

7.5 项目练习

1. 设计一款 App,播放保存在手机 SD 卡中的一个音频文件。
2. 设计一款 App,实现对一个音频文件音量的实时调控。
3. 设计一款 App,实现播放、暂停、停止手机 SD 卡中的一个视频文件。
4. 设计一款 App,实现实时拍照并保存的功能。

项目8

设置App的图像与动画

8.1 项目目标：为 App 添加或设置图像与动画

App 的功能实现需要通过文字、图像、音频、视频以及动画等富媒体形式来充分展现。在前面的项目中，对于如何在 App 中设置及控制文字、音频、视频等多媒体类型已经介绍得较为翔实，本项目将对如何在 App 中设置及控制图像与动画进行实例讲解，为 App 的呈现锦上添花。

8.2 项目准备

8.2.1 介绍绘图类

App 界面上的图像可以通过多种途径实现，一种途径是通过设置 Image 相关类进行已有图像的引用，例如 App 的背景图像、按钮的背景图像等；另一种途径是通过绘制的方式在 App 界面上直接由代码实现。使用绘图的方式制作图像相较于引用图像的方式，可能在图像的成像水平、画面精美程度等方面并不占有优势，但这是一种较为自由、互动性强的方法，因此有学习的必要。

在 Android 开发中，可以通过 Paint（画笔）类、Canvas（画布）类、Bitmap（位图）类等实现绘图。

1. Paint 类

Paint 即画笔，当 Paint 在 Canvas（即画布）上绘制时，即可生成图像。与任何一款绘制图像的工具类相比，画笔能够帮助绘画者实现多种效果与风格，例如颜色、透明度、字体风格、抗锯齿、笔刷宽度、图像渐变效果等。在构建 Paint 类时多采用 new 的方法，具体代码如下：

```
Paint paint = new Paint();
```

为了实现各种画笔效果，在 Android 中有多种方法可供 Paint 类调用，具体方法名称和使用说明如表 8-1 所示。

表 8-1 Paint 类常用的方法和使用说明

方 法 名 称	使 用 说 明
setARGB(a,r,g,b)	设置画笔的颜色。其 4 个参数分别代表透明度和颜色的 RGB 值，取值范围为 0～255
setAlpha(a)	设置画笔的 Alpha 值。其范围为 0～255，0 代表完全透明，255 代表完全不透明
setAntiAlias(true)	设置画笔的锯齿效果。true 代表抗锯齿，false 代表不抗锯齿
setColor(R.color.red)	设置画笔的颜色。其参数为 int 类型
setColorFilter(ColorFilter colorfilter)	设置颜色过滤器，可以在绘制颜色时实现不用颜色的变换效果
setDither(true)	设置是否使用图像抖动处理，使绘制出来的图片的颜色更加平滑和饱满，图像更加清晰
setFakeBoldText(true)	模拟实现粗体文字，设置在小字体上效果会非常差
setFilterBitmap(true)	如果该项设置为 true，则图像在动画进行中会滤掉对 Bitmap 图像的优化操作，加快显示速度。本设置项依赖于 dither 和 xfermode 的设置
setFlags(Paint.ANTI_ALIAS_FLAG)	根据 flag 值对画笔进行设置。例如这里设置的是抗锯齿

续表

方法名称	使用说明
setMaskFilter(MaskFilter maskfilter)	设置MaskFilter，可以用不同的MaskFilter实现滤镜的效果，例如滤化、立体等
setPathEffect(PathEffect effect)	设置绘制路径的效果，例如点画线等
setShader(Shader shader)	设置图像效果，使用Shader可以绘制出各种渐变效果
setShadowLayer(float radius, float dx, float dy, int color)	在图形下面设置阴影层，产生阴影效果。radius为阴影的角度，dx和dy为阴影在X轴和Y轴上的距离，color为阴影的颜色
setStrikeThruText(boolean strikeThruText)	设置带有删除线的效果
setStrokeCap(Paint.Cap.ROUND)	当画笔样式为STROKE或FILL_OR_STROKE时，设置笔刷的图形样式，例如圆角形样式(Cap.ROUND)或方形样式(Cap.SQUARE)。这会影响画笔的始末端
setSrokeJoin(Paint.Join join)	设置绘制时各图形的结合方式，例如平滑效果等
setStrokeWidth(float width)	当画笔样式为STROKE或FILL_OR_STROKE时，设置笔刷的粗细度，即宽度
setStyle(Paint.Style style)	设置画笔的样式，包括FILL(实心的)、FILL_OR_STROKE或STROKE(空心的)
setSubpixelText(boolean subpixelText)	设置该项为true，将有助于文本在LCD屏幕上的显示效果
setTextAlign(Paint.Align align)	设置绘制文字的对齐方向
setTextScaleX(float scaleX)	设置绘制文字在X轴上的缩放比例，可以实现文字的拉伸效果
setTextSize(float textSize)	设置绘制文字的字号大小
setTextSkewX(float skewX)	设置斜体文字，skewX为倾斜弧度
setTypeface(Typeface typeface)	设置Typeface对象，即字体风格，包括粗体、斜体以及衬线体、非衬线体等
setUnderlineText(boolean underlineText)	设置带有下画线的文字效果

2. Canvas类

Canvas即画布，所有的图形图像都要在画布上成像，开发者可以通过API中提供的方法在Canvas上进行绘制，例如绘制各种形状等。在创建Canvas时，可以通过两种方式实现绘图功能：一种是通过View类创建对象；另一种是使用SurfaceView组件下的Canvas类创建对象。两种方法适用的情境不同，对于体量小、帧频低的图像或动画可以直接使用第一种方法；对于有高品质要求的图像或动画而言，第二种方法更为适用。在Android中，Canvas类常用的方法如表8-2所示。

表 8-2 Canvas 类常用的方法及使用说明

方法名称	使用说明
drawRect(floot left,floot top,floot right,floot bottom,Paint paint)	绘制一个矩形,参数 1 为矩形左侧边距离原点的距离,参数 2 为矩形上边距离原点的距离,参数 3 为矩形右侧边距离原点的距离,参数 4 为矩形下边距离原点的距离,参数 5 为 Paint 对象
drawPath(Path path,Paint paint)	绘制一个路径,参数 1 为 Path 对象
drawBitmap(Bitmap bitmap,Rect src,Rect dst,Paint paint)	贴图,参数 1 就是常规的 Bitmap 对象,参数 2 是源区域(这里是 bitmap),参数 3 是目标区域(应该为 Canvas 的位置和大小),参数 4 是 Paint(画刷)对象,因为有缩放和拉伸的可能,当原始 Rect 不等于目标 Rect 时性能会有大幅损失
drawLine(float startX,float startY,float stopX,float stopY,Paintpaint)	画线,参数 1 为起始点的 X 轴位置,参数 2 为起始点的 Y 轴位置,参数 3 为终点的 X 轴水平位置,参数 4 为终点的 Y 轴垂直位置,最后一个参数为 Paint(画刷)对象
drawPoint(float x,float y,Paint paint)	画点,参数 1 为水平 X 轴,参数 2 为垂直 Y 轴,参数 3 为 Paint 对象
drawText(String text,float x,float y,Paint paint)	渲染文本,参数 1 是 String 类型的文本,参数 2 是 X 轴,参数 3 是 Y 轴,参数 4 是 Paint 对象
drawOval(RectF oval,Paint paint)	画椭圆,参数 1 是扫描区域,参数 2 是 Paint 对象
drawCircle(float cx,float cy,float radius,Paint paint)	画圆,参数 1 是中心点的 X 轴,参数 2 是中心点的 Y 轴,参数 3 是半径,参数 4 是 Paint 对象
drawArc(RectF oval,float startAngle,float sweepAngle,boolean useCenter,Paint paint)	画弧,参数 1 是 RectF 对象,一个矩形区域椭圆形的界限用于定义形状、大小、电弧;参数 2 是起始角(度)在电弧的开始;参数 3 扫描角(度)开始顺时针测量,参数 4 如果为真,包括椭圆中心的电弧,并关闭它,如果为假,将是一个弧线;参数 5 是 Paint 对象

3. Bitmap 类

Bitmap 即位图,在 Android 开发中它是图像处理最重要的类之一。使用该类可以获取图像文件信息,对图像进行剪切、旋转、缩放等操作,并可以指定格式保存图像文件。为了实现这些操作,Bitmap 类支持表 8-3 中的方法。

表 8-3 Bitmap 类支持的常用方法及使用说明

方法名称	使用说明
recycle()	回收位图占用的内存空间,把位图标记为 Dead
boolean isRecycled()	判断位图内存是否已释放

续表

方法名称	使用说明
getWidth()	获取位图的宽度
getHeight()	获取位图的高度
isMutable()	图片是否可修改
getScaledWidth(Canvas canvas)	获取指定密度转换后的图像的宽度
getScaledHeight(Canvas canvas)	获取指定密度转换后的图像的高度
compress(CompressFormat format,int quality,OutputStream stream)	按指定的图片格式以及画质将图片转换为输出流
createBitmap(Bitmap src)	以 src 为原图生成不可变的新图像
createScaledBitmap(Bitmap src,int dstWidth,int dstHeight,boolean filter)	以 src 为原图创建新的图像,指定新图像的高、宽以及是否可变
createBitmap(int width,int height,Config config)	创建指定格式、大小的位图
createBitmap(Bitmap source,int x,int y,int width,int height)	以 source 为原图创建新的图片,指定起始坐标以及新图像的高、宽

8.2.2　介绍图像特效

在 Android 中可以对图像做进一步处理,包括设置其旋转、缩放、倾斜、平移等,这些效果统称为图像特效。其中,Rotate 类用于实现图像旋转;Scale 类用于实现图像缩放;Matrix 类用于实现图像倾斜;Translate 类用于实现图像平移等。需要注意的是,Matrix 类的作用对象是 Bitmap,而不是 Canvas。

1. Rotate 类实现图像旋转

Rotate(旋转变换)采用一个浮点数表示旋转的角度。通常,围绕默认点(0,0),正数将顺时针旋转图像,负数将逆时针旋转图像,其中默认点是图像的左上角。

```
Matrix matrix = new Matrix();
matrix.setRotate(60);
canvas.drawBitmap(bmp, matrix, paint);
```

当然,也可以使用旋转的角度及围绕的旋转点作为参数调用 setRotate()方法。这种方式可以选择图像的中心点作为旋转点。

```
matrix.setRotate(30, bmp.getWidth()/2, bmp.getHeight()/2);
```

2. Scale 类实现图像缩放

Scale(缩放变换)采用两个浮点数作为参数,分别表示在 X 轴和 Y 轴上所产生的缩放量。第一个参数是 X 轴的缩放比例,第二个参数是 Y 轴的缩放比例。

```
matrix.setScale(2f, 1.5);
```

3. Translate 类实现平移

Translate(平移变换)采用两个浮点数作为参数,表示在 X 轴和 Y 轴上移动的数量。第一个参数是图像在 X 轴上移动的数量;第二个参数是图像在 Y 轴上移动的数量。在 X 轴上使使用正数进行平移将向右移动图像,使用负数进行平移将向左移动图像;在 Y 轴上使用正数进行平移将向下移动图像,使用负数进行平移将向上移动图像。

```
matrix.setTranslate(2f, -5);
```

8.2.3 介绍动画类型

在设计开发 App 时,动画的使用是十分普遍的。有动画形式的加入,能够使得 App 更具动感与趣味性。与动画的制作原理相似,在 Android 开发中对于动画的实现一般也是通过两种基本方式,即属性动画(Property Animation)和视图动画(View Animation),而视图动画又包含两种类型,一种是源于动画的"视觉暂留"原理而提出的"逐帧动画",一种是基于计算机辅助生成原理而提出的"补间动画"。两种动画形式对于资源的需求不同,方法的使用不同,呈现的效果也有所区别。

在 Android 中,动画的实现可以由 Animation 类来完成,它的 XML 属性如表 8-4 所示。

表 8-4 Animation 类的 XML 属性

属性名称	属性含义
android:detachWallpaper	设置是否让桌面壁纸跟着动画。其默认是 false
android:duration	设置动画执行的时间
android:fillAfter	动画结束时停留在最后一帧
android:fillBefore	动画结束时停留在第一帧
android:fillEnabled	与 fillBefore 结合使用
android:interpolator	设置动画为加速动画

续表

属 性 名 称	属 性 含 义
android：repeatCount	设置动画重复的次数
android：repeatMode	设置动画重复的方式
android：startOffset	设置动画延迟执行的时间
android：zAdjustment	被设置动画的内容在动画运行时在 Z 轴上的位置

1. 逐帧动画

逐帧动画的原理是将每一张图片作为一帧，以预先设定的顺序进行播放，即形成动画的效果。实现逐帧动画的具体步骤如下。

步骤 1：将动画资源（即图片序列）放到 res 下的 drawable 文件夹中等待使用，如图 8-1 所示。

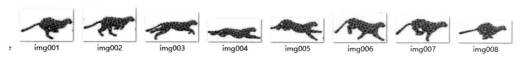

图 8-1　逐帧动画图片序列

步骤 2：设置动画，即在 res 文件夹中新建一个 XML 文件，并在其中定义动画的图片资源。其具体代码如下：

```xml
<animation-list
    xmlns:android="http://schemas.android.com/apk/res/android"
    android:oneshot="true">
    <item android:drawable="@drawable/img001" android:duration="100"/>
    ...
    <item android:drawable="@drawable/img00N" android:duration="100"/>
</animation-list>
```

步骤 3：启动动画，即在 MainActivity.java 文件中设置动画的播放等控制方法。其具体代码如下：

```java
frameanimi.setOnClickListener(new View.OnClickListener() {
    @Override
    public void onClick(View v) {
        animationDrawable = (AnimationDrawable) v.getDrawable();
        animationDrawable.start();
    }
}
```

2. 补间动画

补间动画的形成与逐帧动画不同,它是依托于计算机辅助生成的方法实现,补间动画与关键帧密不可分。所谓关键帧,即动画形成的重要帧幅。补间动画是计算机通过两个相邻的关键帧之间的细微变化进行补充绘制所形成的画面。一般而言,在 Android 开发中补间动画有旋转、缩放、位移、透明度等效果的绘制。

1) RotateAnimation(旋转补间动画)

RotateAnimation 类是 Android 系统中的旋转变化动画类,用于控制 View 对象的旋转动作,该类继承于 Animation 类。RotateAnimation 类中的很多方法都与 Animation 类一致,该类中最常用的方法就是 RotateAnimation 的构造方法。其实现的基本语法如下:

```
public RotateAnimation(float fromDegrees, float toDegrees, int pivotXType, float pivotXValue,
int pivotYType, float pivotYValue)
```

其中,fromDegrees 代表旋转的开始角度;toDegrees 代表旋转的结束角度;pivotXType 代表 X 轴的伸缩模式,可以取值为 ABSOLUTE、RELATIVE_TO_SELF、RELATIVE_TO_PARENT;pivotXValue 代表 X 轴坐标的伸缩值;pivotYType 代表 Y 轴的伸缩模式,可以取值为 ABSOLUTE、RELATIVE_TO_SELF、RELATIVE_TO_PARENT;pivotYValue 代表 Y 轴坐标的伸缩值。

2) ScaleAnimation(缩放补间动画)

ScaleAnimation 类是 Android 系统中的缩放变化动画类,用于控制 View 对象的缩放动作,该类继承于 Animation 类。与 RotateAnimation 类似,该类中最常用的方法就是 ScaleAnimation 的构造方法。其实现的基本语法如下:

```
public ScaleAnimation(float fromX, float toX, float fromY, float toY, int pivotXType, float
pivotXValue, int pivotYType, float pivotYValue)
```

其中,float fromX 代表动画起始时 X 轴坐标上的伸缩尺寸;float toX 代表动画结束时 X 轴坐标上的伸缩尺寸;float fromY 代表动画起始时 Y 轴坐标上的伸缩尺寸;float toY 代表动画结束时 Y 轴坐标上的伸缩尺寸;int pivotXType 代表动画在 X 轴相对于物件位置类型;float pivotXValue 代表动画相对于物件的 X 轴坐标的开始位置;int pivotYType 代表动画在 Y 轴相对于物件位置类型;float pivotYValue 代表动画相对于物件的 Y 轴坐标的开始位置。

3）TranslateAnimation（位移补间动画）

TranslateAnimation 类是 Android 系统中的位移变化动画类，用于控制 View 对象的位移动作，该类继承于 Animation 类。与 RotateAnimation 类似，该类中最常用的方法就是 TranslateAnimation 的构造方法。其实现的基本语法如下：

```
TranslateAnimation(float fromXDelta, float toXDelta, float fromYDelta, float toYDelta)
```

其中，fromXDelta 代表动画开始点的 X 轴坐标；fromYDelta 代表动画开始点的 Y 轴坐标；toXDelta 代表动画结束点的 X 轴坐标；toYDelta 代表动画结束点的 Y 轴坐标。

4）AlphaAnimation（透明度补间动画）

AlphaAnimation 类是 Android 系统中的透明度变化动画类，用于控制 View 对象的透明度变化，该类继承于 Animation 类。与 RotateAnimation 类似，该类中最常用的方法就是 AlphaAnimation 的构造方法。其实现的基本语法如下：

```
public AlphaAnimation(float fromAlpha, float toAlpha)
```

其中，fromAlpha 代表开始时刻的透明度，取值范围为 0～1；toAlpha 代表结束时刻的透明度，取值范围为 0～1。

8.3 项目运行

8.3.1 添加图形图像

【例 8.1】 在 Android Studio 中创建名为 8.1 的 App 项目，使用 Paint 类定义 3 种不同类型的画笔，并用其绘制出渐变类型不同的 3 个矩形。

步骤 1：在 MainActivity.java 文件中对第一支画笔进行定义，并为矩形填充线性渐变颜色。其具体代码如下：

视频讲解

```
protected void onDraw(Canvas canvas) {
    Paint paint1 = new Paint();
    Shader shader1 = new LinearGradient(0, 0, 50, 50, Color.RED, Color.GREEN, Shader.TileMode.MIRROR);
```

```
        paint.setShader(shader1);
        canvas.drawRect(120,200,220,280, paint1);
         super.onDraw(canvas);
}
```

显示效果如图 8-2 所示。

图 8-2　线性渐变效果

步骤 2：在 MainActivity.java 文件中对第二支画笔进行定义，并为矩形填充角度渐变颜色。其具体代码如下：

```
protected void onDraw(Canvas canvas) {
    Paint paint2 = new Paint();
    Shader shader2 = new SweepGradient(265,110,new int[]
                {Color.RED,Color.GREEN,Color.BLUE},null);
    Paint2.setShader(shader2);
    canvas.drawRect(120,200,220,280,paint);
     super.onDraw(canvas);
}
```

显示效果如图 8-3 所示。

步骤 3：在 MainActivity.java 文件中对第三支画笔进行定义，并为矩形填充径向渐变颜色。其具体代码如下：

```
protected void onDraw(Canvas canvas) {
        Paint paint3 = new Paint();
        Shader shader3 = new RadialGradient(160, 110, 50, Color.RED, Color.GREEN,
                    Shader.TileMode.MIRROR);
        Paint3.setShader(shader3);           //为画笔设置渐变器
        canvas.drawRect(120,200,220,280,paint);
         super.onDraw(canvas);
}
```

步骤 4：运行程序，效果如图 8-4 所示。

图 8-3　角度渐变效果

图 8-4　径向渐变效果

【例 8.2】　在 Android Studio 中创建名为 8.2 的 App 项目，使用 Canvas 类实现画布绘图功能。

步骤 1：在 Android Studio 中创建一个名为 8.2 的 App 项目，通过继承

视频讲解

View 类的方式创建名为 CanvasView 的新类,然后在 src 的 main 下找到 Java 中的 com 文件夹,新建 CanvasView.java 文件,进行画布的创建。其具体代码如下:

```java
package com.example.helloworld;
import android.content.Context;
import android.graphics.Canvas;
import android.graphics.Color;
import android.graphics.Paint;
import android.support.annotation.Nullable;
import android.util.AttributeSet;
import android.view.View;

public class CanvasView extends View {
    public CanvasView (Context context, AttributeSet attrs) {
        super(context, attrs);
    }
}
```

步骤 2:在新建的 CanvasView(画布)上绘制一个蓝色的矩形。其具体代码如下:

```java
@Override
    protected void onDraw(Canvas canvas) {
        Paint paint = new Paint();
        paint.setColor(Color.BLUE);
        canvas.drawRect(45, 45, 450, 250, paint);
        super.onDraw(canvas);
    }
```

步骤 3:在项目文件中找到 res 下 layout 文件夹中的 activity_main.xml 文件,设置布局管理器。其具体代码如下:

```xml
<?xml version = "1.0" encoding = "utf-8"?>
< FrameLayout xmlns:android = "http://schemas.android.com/apk/res/android"
    android:layout_width = "fill_parent"
    android:layout_height = "fill_parent"
    >
    < com.example.helloworld.CanvasView
        android:id = "@ + id/canvasView1"
        android:layout_width = "wrap_content"
        android:layout_height = "wrap_content" />
</FrameLayout >
```

步骤 4:运行程序,效果如图 8-5 所示。

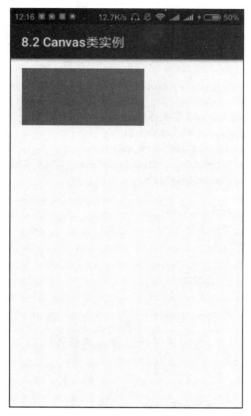

图 8-5　Canvas 类绘制矩形的运行界面

8.3.2　设计图像特效

【例 8.3】　在 Android Studio 中创建名为 8.3 的 App 项目,使用 Matrix 类实现图像的旋转。

步骤 1:在 Android Studio 中创建一个名为 8.3 的 App 项目,并在项目文件中找到 res 下 layout 文件夹中的 activity_main.xml 文件,添加一个 FrameLayout 组件用于显示图像。其具体代码如下:

视频讲解

```xml
<?xml version = "1.0" encoding = "utf-8"?>
<FrameLayout xmlns:android = "http://schemas.android.com/apk/res/android"
    android:id = "@ + id/frameLayout1"
    android:layout_width = "fill_parent"
    android:layout_height = "fill_parent"
    android:orientation = "vertical" >
</FrameLayout>
```

步骤2：在MainActivity.java文件中创建一个View类，并将其添加到FrameLayout管理器中。其关键代码如下：

```java
public class MainActivity extends AppCompatActivity {
    @Override
    public void onCreate(Bundle savedInstanceState) {
        super.onCreate(savedInstanceState);
        setContentView(R.layout.activity_main);
        FrameLayout fl = (FrameLayout)findViewById(R.id.frameLayout1);
        fl.addView(new MyIamge(this));
    }
    public class MyIamge extends View{
        public MyIamge(Context context) {
            super(context);
        }
    }
}
```

步骤3：在MainActivity.java文件中定义一个画笔，通过drawBitmap()方法绘制一张图片，这里使用名为pic的图片，再将其进行旋转。其具体代码如下：

```java
@Override
protected void onDraw(Canvas canvas) {
    Paint paint = new Paint();
    paint.setAntiAlias(true);
    Bitmap bm_pic = BitmapFactory.decodeResource(MainActivity.this.getResources(),
        R.drawable.pic);
    canvas.drawBitmap(bm_pic, 0, 0, paint);            // 绘制原图
    Matrix m1 = new Matrix();
    matrix.setRotate(25);                              // 将原图以图像轴心为原点旋转25°
    canvas.drawBitmap(bm_pic, matrix, paint);
    Matrix m2 = new Matrix();
    m.setRotate(90,50,50);                             // 将原图以(50,50)为原点旋转90°
    canvas.drawBitmap(bitmap_pic, m, paint);
    super.onDraw(canvas);
}
```

步骤4：运行程序，效果如图8-6所示。

【例8.4】 在Android Studio中创建名为8.4的App项目，使用Matrix类实现图像的缩放。

步骤1：在Android Studio中创建一个名为8.4的App项目，并在项目文件中找到res下layout文件夹中的activity_main.xml文件，添加一个FrameLayout组件用于显示图像。其具体代码如下：

视频讲解

项目8　设置App的图像与动画

图 8-6　Matrix 类旋转图像的运行界面

```xml
<?xml version = "1.0" encoding = "utf-8"?>
< FrameLayout xmlns:android = "http://schemas.android.com/apk/res/android"
    android:id = "@ + id/frameLayout1"
    android:layout_width = "fill_parent"
    android:layout_height = "fill_parent"
    android:orientation = "vertical" >
</FrameLayout >
```

步骤 2：在 MainActivity.java 文件中创建一个 View 类，并将其添加到 FrameLayout 管理器中。其关键代码如下：

```java
public class MainActivity extends AppCompatActivity {
    @Override
    public void onCreate(Bundle savedInstanceState) {
        super.onCreate(savedInstanceState);
        setContentView(R.layout.activity_main);
        FrameLayout fl = (FrameLayout)findViewById(R.id.frameLayout1);
        fl.addView(new MyIamge (this));
    }
    public class MyIamge extends View{

        public MyIamge (Context context) {
```

```
        super(context);
    }
}
```

步骤3：在MainActivity.java文件中定义一个画笔，通过drawBitmap()方法绘制一张图片，这里使用名为pic的图片，再将其进行缩放。其具体代码如下：

```
@Override
protected void onDraw(Canvas canvas) {
    Paint paint = new Paint();
    paint.setAntiAlias(true);
    Bitmap bm_pic = BitmapFactory.decodeResource(MainActivity.this.getResources(),
            R.drawable.pic);
    canvas.drawBitmap(bm_pic, 0, 0, paint);
    Matrix m1 = new Matrix();
    m1.setScale(2f,2f);
    canvas.drawBitmap(bm_pic, matrix, paint);
    super.onDraw(canvas);
}
```

步骤4：运行程序，效果如图8-7所示。

图8-7　Matrix类缩放图像的运行界面

【例 8.5】 在 Android Studio 中创建名为 8.5 的 App 项目,使用 Matrix 类实现图像的倾斜。

视频讲解

步骤 1:在 Android Studio 中创建一个名为 8.5 的 App 项目,并在项目文件中找到 res 下 layout 文件夹中的 activity_main.xml 文件,添加一个 FrameLayout 组件用于显示图像。其具体代码如下:

```xml
<?xml version = "1.0" encoding = "utf-8"?>
<FrameLayout xmlns:android = "http://schemas.android.com/apk/res/android"
    android:id = "@+id/frameLayout1"
    android:layout_width = "fill_parent"
    android:layout_height = "fill_parent"
    android:orientation = "vertical" >
</FrameLayout>
```

步骤 2:在 MainActivity.java 文件中创建一个 View 类,并将其添加到 FrameLayout 管理器中。其关键代码如下:

```java
public class MainActivity extends AppCompatActivity {
    @Override
    public void onCreate(Bundle savedInstanceState) {
        super.onCreate(savedInstanceState);
        setContentView(R.layout.activity_main);
        FrameLayout fl = (FrameLayout)findViewById(R.id.frameLayout1);
        fl.addView(new MyIamge (this));
    }
    public class MyIamge extends View{

        public MyIamge (Context context) {
            super(context);
        }
}}
```

步骤 3:在 MainActivity.java 文件中定义一个画笔,通过 drawBitmap()方法绘制一张图片,这里使用名为 pic 的图片,再将其进行倾斜。其具体代码如下:

```java
        @Override
        protected void onDraw(Canvas canvas) {
            Paint paint = new Paint();
            paint.setAntiAlias(true);
Bitmap bp_pic = BitmapFactory.decodeResource(MainActivity.this.getResources(), R.drawable.pic);
            Matrix m1 = new Matrix();
            m1.setSkew(2f, 1f);
            canvas.drawBitmap(bp_pic, matrix, paint);
```

```
            canvas.drawBitmap(bp_pic, 0, 0, paint);
            super.onDraw(canvas);          ──→ 绘制原图
        }
    }
```

步骤4:运行程序,效果如图8-8所示。

图8-8 Matrix类倾斜图像的运行界面

【例8.6】 在Android Studio中创建名为8.6的App项目,使用Matrix类实现图像的平移。

视频讲解

步骤1:在Android Studio中创建一个名为8.6的App项目,并在项目文件中找到res下layout文件夹中的activity_main.xml文件,添加一个FrameLayout组件用于显示图像。其具体代码如下:

```
<?xml version = "1.0" encoding = "utf-8"?>
< FrameLayout xmlns:android = "http://schemas.android.com/apk/res/android"
    android:id = "@ + id/frameLayout1"
    android:layout_width = "fill_parent"
    android:layout_height = "fill_parent"
```

```
        android:orientation = "vertical" >
</FrameLayout >
```

步骤 2：在 MainActivity.java 文件中创建一个 View 类,并将其添加到 FrameLayout 管理器中,关键代码如下：

```
public class MainActivity extends AppCompatActivity {
    @Override
    public void onCreate(Bundle savedInstanceState) {
        super.onCreate(savedInstanceState);
        setContentView(R.layout.activity_main);
        FrameLayout fl = (FrameLayout)findViewById(R.id.frameLayout1);
        fl.addView(new MyIamge (this));
    }
    public class MyIamge extends View{

        public MyIamge (Context context) {
            super(context);
        }
}}
```

步骤 3：在 MainActivity.java 文件中定义一个画笔,通过 drawBitmap()方法绘制一张图片,这里使用名为 pic 的图片,再将其进行平移。其具体代码如下：

```
        @Override
        protected void onDraw(Canvas canvas) {
            Paint paint = new Paint();
            paint.setAntiAlias(true);
Bitmap bm_pic = BitmapFactory.decodeResource(MainActivity.this.getResources(),
                R.drawable.pic);
            Matrix m1 = new Matrix();
            m1.postTranslate(50, 200);
            canvas.drawBitmap(bm_pic, matrix, paint);
            canvas.drawBitmap(bm_pic, 0, 0, paint);      ──→ 绘制原图
            super.onDraw(canvas);
        }
    }
}
```

步骤 4：运行程序,效果如图 8-9 所示。

8.3.3 设计动画

【例 8.7】 在 Android Studio 中创建名为 8.7 的 App 项目,实现逐帧动

视频讲解

图 8-9 Matrix 类平移图像的运行界面

画的生成。

步骤 1：在 Android Studio 中创建一个名为 8.7 的 App 项目，首先将动画资源（img001～img008）放到 res 下的 drawable-v24 文件夹中。

步骤 2：在 res 文件夹中新建一个 drawable 文件夹，在其中添加一个名为 leopard 的 XML 文件，并在里面定义动画图片序列。其关键代码如下：

```xml
<animation-list
    xmlns:android="http://schemas.android.com/apk/res/android"
    android:oneshot="true">
    <item android:drawable="@drawable/img001" android:duration="100"/>
    <item android:drawable="@drawable/img002" android:duration="100"/>
    <item android:drawable="@drawable/img003" android:duration="100"/>
    <item android:drawable="@drawable/img004" android:duration="100"/>
    <item android:drawable="@drawable/img005" android:duration="100"/>
    <item android:drawable="@drawable/img006" android:duration="100"/>
    <item android:drawable="@drawable/img007" android:duration="100"/>
    <item android:drawable="@drawable/img008" android:duration="100"/>
</animation-list>
```

步骤3：在项目文件中找到res下layout文件夹中的activity_main.xml文件，创建逐帧动画资源，并进行布局的设置。其具体代码如下：

```xml
<?xml version = "1.0" encoding = "utf-8"?>
<LinearLayout xmlns:android = "http://schemas.android.com/apk/res/android"
    android:layout_gravity = "center"
    android:layout_width = "match_parent"
    android:layout_height = "wrap_content"
    android:background = "@drawable/leopard"
    android:id = "@ + id/ll"
>
</LinearLayout>
```

步骤4：在MainActivity.java文件中实现逐帧动画的播放。其具体代码如下：

```java
package com.example.helloworld;

import android.graphics.drawable.AnimationDrawable;
import android.os.Bundle;
import android.support.v7.app.AppCompatActivity;
import android.view.View;
import android.view.View.OnClickListener;
import android.widget.LinearLayout;

public class MainActivity extends AppCompatActivity {
    private boolean flag = true;
    @Override
    public void onCreate(Bundle savedInstanceState) {
        super.onCreate(savedInstanceState);
        setContentView(R.layout.activity_main);
        LinearLayout ll = (LinearLayout)findViewById(R.id.ll);
        final AnimationDrawable drawable = (AnimationDrawable)ll.getBackground();
        ll.setOnClickListener(new OnClickListener() {
            @Override
            public void onClick(View v) {
                animationDrawable = (AnimationDrawable) v.getDrawable();
                animationDrawable.start();
            }
        }
    }
}
```

步骤5：运行程序，效果如图8-10所示。

【例8.8】 在Android Studio中创建名为8.8的App项目，实现补间动画效果的生成，包括旋转、缩放、位移、透明度等效果。

视频讲解

图 8-10 逐帧动画的运行界面

步骤1：在 Android Studio 中创建一个名为 8.8 的 App 项目，首先在 res 文件夹下创建一个用于实现补间动画效果的动画资源文件夹，名为 animation。

步骤2：在 animation 文件夹下创建用于实现旋转动画的 XML 文件，名为 animation_rotate.xml。其具体代码如下：

```xml
<?xml version = "1.0" encoding = "utf-8"?>
<set xmlns:android = "http://schemas.android.com/apk/res/android">
    <rotate
        android:interpolator = "@android:animation/accelerate_interpolator"
        android:fromDegrees = "0"
        android:toDegrees = "680"
        android:pivotX = "50%"
        android:pivotY = "50%"
        android:duration = "3000">
    </rotate>
    <rotate
        android:interpolator = "@android:anim/accelerate_interpolator"
        android:startOffset = "3000"
        android:fromDegrees = "340"
```

```
            android:toDegrees = "0"
            android:pivotX = "50%"
            android:pivotY = "50%"
            android:duration = "3000">
        </rotate>
</set>
```

步骤3：在animation文件夹下创建用于实现缩放动画的XML文件，名为animation_scale.xml。其具体代码如下：

```
<?xml version = "1.0" encoding = "utf-8"?>
<set xmlns:android = "http://schemas.android.com/apk/res/android">
    <scale android:fromXScale = "1"
        android:interpolator = "@android:animation/decelerate_interpolator"
        android:fromYScale = "1"
        android:toXScale = "2.0"
        android:toYScale = "2.0"
        android:pivotX = "60%"
        android:pivotY = "60%"
        android:fillAfter = "true"
        android:repeatCount = "1"
        android:repeatMode = "reverse"
        android:duration = "3000"/>
</set>
```

步骤4：在animation文件夹下创建用于实现平移动画的XML文件，名为animation_translate.xml。其具体代码如下：

```
<?xml version = "1.0" encoding = "utf-8"?>
<set xmlns:android = "http://schemas.android.com/apk/res/android">
<translate
    android:fromXDelta = "0"
    android:toXDelta = "280"
    android:fromYDelta = "0"
    android:toYDelta = "0"
    android:fillAfter = "true"
    android:repeatMode = "reverse"
    android:repeatCount = "1"
    android:duration = "3000">
</translate>
</set>
```

步骤5：在animation文件夹下创建用于实现动画透明度的XML文件，名为animation_alpha.xml。其具体代码如下：

```xml
<?xml version = "1.0" encoding = "utf-8"?>
<set xmlns:android = "http://schemas.android.com/apk/res/android">
    <alpha android:fromAlpha = "1"
        android:toAlpha = "0"
        android:fillAfter = "true"
        android:repeatMode = "reverse"
        android:repeatCount = "1"
        android:duration = "3000"/>
</set>
```

步骤6:为了实现补间动画的效果,需要在 res 下 layout 文件夹中的 activity_main.xml 文件中布局相关图像资源,并添加执行按钮。其具体代码如下:

```xml
<?xml version = "1.0" encoding = "utf-8"?>
<LinearLayout xmlns:android = "http://schemas.android.com/apk/res/android"
    android:id = "@+id/linearLayout1"
    android:layout_width = "fill_parent"
    android:layout_height = "fill_parent"
    android:orientation = "vertical" >
    <LinearLayout
        android:id = "@+id/linearLayout2"
        android:layout_width = "match_parent"
        android:layout_height = "wrap_content"
        android:orientation = "horizontal" >
        <Button
            android:id = "@+id/button_rotate"
            android:layout_width = "wrap_content"
            android:layout_height = "wrap_content"
            android:text = "Rotate" />         ← 添加执行旋转动画的按钮
        <Button
            android:id = "@+id/button_translate"
            android:layout_width = "wrap_content"
            android:layout_height = "wrap_content"
            android:text = "Translate" />      ← 添加执行平移动画的按钮

        <Button
            android:id = "@+id/button_scale"
            android:layout_width = "wrap_content"
            android:layout_height = "wrap_content"
            android:text = "Scale" />          ← 添加执行缩放动画的按钮

        <Button
            android:id = "@+id/button_alpha"
            android:layout_width = "wrap_content"
            android:layout_height = "wrap_content"
            android:text = "Alpha" />          ← 添加执行透明度动画的按钮
```

```
        </LinearLayout >

        < ImageView
            android:id = "@ + id/imageView1"
            android:layout_width = "wrap_content"
            android:layout_height = "wrap_content"
            android:layout_marginLeft = "50px"
            android:src = "@drawable_v24/car1" />
</LinearLayout >
```

步骤7:为了获取实现补间动画的图像资源,需要在 MainActivity.java 文件中编写获取资源的代码,并为按钮添加事件。其具体代码如下:

```
public class MainActivity extends AppCompatActivity {
    @Override
    public void onCreate(Bundle savedInstanceState) {
        super.onCreate(savedInstanceState);
        setContentView(R.layout.activity_main);
        final Animation rotate = AnimationUtils.loadAnimation(this, R.animition.animition_rotate);
        final Animation translate = AnimationUtils.loadAnimation(this, R.animition.animition_translate);
        final Animation scale = AnimationUtils.loadAnimation(this, R.animition.animition_scale);
        final Animation alpha = AnimationUtils.loadAnimation(this, R.animition.animition_alpha);
        final ImageView pic = (ImageView)findViewById(R.id.imageView1);
        Button button_rotate = (Button)findViewById(R.id.button_rotate);
        button_rotate.setOnClickListener(new OnClickListener() {
            @Override
            public void onClick(View v) {
                pic.startAnimation(rotate);
            }
        });
        Button button_translate = (Button)findViewById(R.id.button_translate);
        button_translate.setOnClickListener(new OnClickListener() {
            @Override
            public void onClick(View v) {
                pic.startAnimation(translate);
            }
        });
        Button button_scale = (Button)findViewById(R.id.button_scale);
        button_scale.setOnClickListener(new OnClickListener() {
            @Override
            public void onClick(View v) {
                pic.startAnimation(scale);
            }
        });
```

获取animation文件夹下的4个补间动画资源

旋转按钮事件

```
Button button_alpha = (Button)findViewById(R.id.button_alpha);
button_alpha.setOnClickListener(new OnClickListener() {

    @Override
    public void onClick(View v) {
        pic.startAnimation(alpha);
    }
});
}
```

步骤 8：运行程序，效果如图 8-11 所示。

图 8-11　补间动画的运行界面

8.4　项目结案

本项目通过实例的方式为大家介绍了 Android 常用的处理图形图像的方法，包括图像的绘制、设置图像的特效、动画的实现等。在图像的绘制中向大家介绍了 Paint（画笔）类、

Canvas(画布)类、Bitmap(位图)类等的运用与设置,这些方法尤其适用于交互性强的 App 的开发与设计;在设置图像的特效中向大家介绍了图像的旋转、平移、旋转等效果;在动画的实现中介绍了逐帧动画和补间动画的区别以及设置的方式。本项目对于在 App 中实现图形图像功能而言是全面且实用的,希望大家能够在练习中体会图形图像与交互之间的关联,让 App 的设计摆脱平面的束缚,更具动态效果与交互体验。

8.5 项目练习

1. 设计一款 App,在界面上绘制两个多边形,且添加不同的颜色。
2. 设计一款 App,利用逐帧动画的序列素材制作一个动画效果,例如野马奔跑。
3. 设计一款 App,实现文字忽明忽暗的效果。
4. 设计一款 App,制作一个交互式动画,实现小球从屏幕的一侧滚动到另一侧。

项目9

获取App的数据

9.1 项目目标：获取数据与线程设置

在使用 App 的过程中，往往存在多个线程同时运行的情况。多线程是相对 UI 主线程而言的，一般来说，当用户操作涉及处理文件的输入/输出，或者网络输入/输出的耗时操作时，多线程异步处理的方法能够避免 UI 线程被阻塞，这样界面才不会无法响应，影响用户体验。

9.2 项目准备

9.2.1 介绍多线程

多线程与人们的现实生活十分相似。例如，人们在日常生活中经常一边听音乐，一边看书，同时做几件事。在使用手机时更是如此，人们经常需要在等待下载的同时再做一些其他事情。为了能够实现这种多线程同时运行且互不影响的效果,在 Android 中提供了多线程机制。一般而言,多线程机制包括创建线程、启动线程、休眠线程和中断线程 4 种状态。

1. 创建线程

在 Android 中可以通过 Runnable 创建线程,首先需要定义类以实现 Runnable 接口;再通过覆盖 Runnable 接口中的 run()方法,将线程要运行的代码存放在该 run()方法中;通过 Thread 类建立线程对象;将 Runnable 接口的子类对象作为实际参数传递给 Thread 类的构造函数;调用 Thread 类的 start()方法开启线程并调用 Runnable 接口子类的 run()方法。其具体代码如下:

```
private Thread newThread;
newThread = new Thread(new Runnable() {
    @Override
    public void run() {
      //线程需要做的工作
        }
    });
```

2. 启动线程

在创建一个线程后,需要启动该线程才能使其运行。启动线程可以使用 start()方法,具体代码如下:

```
newThread.start();
```

3. 休眠线程

休眠指暂停,即使得当前运行的线程暂时停止运行,并当再次启动时恢复至运行状态。在 Android 中为休眠操作提供了同名方法,即 sleep()。其具体代码如下:

```
newThread.sleep(long time);
```

其中,long time 意为休眠时间,单位为毫秒。

4. 中断线程

线程可以被视为一次性消耗品,一般线程在执行完成后便正常结束了。线程结束后不能再次启动，只能新建一个线程对象，但有时 run() 方法是永远不会结束的。例如在程序中使用线程进行 Socket 监听请求，或是其他需要循环处理的任务。在这种情况下，一般是将这些任务放在一个循环中，例如 while 循环。当需要中断线程时有 3 种方法可以选择：一是使用退出标志，使线程正常退出，也就是当 run() 方法完成后线程终止；二是使用 interrupt() 方法中断线程；三是使用 stop() 方法强行终止线程。第三种方法相当于强行关机，不推荐使用。

在一般情况下，推荐开发者使用第二种方法中断线程。其具体代码如下：

```
public class ThreadSafe extends Thread {
    public void run() {
        while (!isInterrupted()){
            …
        }
    }
}
```

9.2.2 介绍消息类

Message 类即为消息类，是线程之间传递信息的载体，其中包含了对消息的描述和数据对象。在 Message 中包含两个额外的 int 字段和一个 object 字段，这样的好处是为开发者省去了内存分配的工作。虽然 Message 的构造函数是 public 的，但最好是使用 Message.obtain() 或 Handler.obtainMessage() 函数来获取 Message 对象，因为 Message 的实现中包含回收再利用的机制，可以提高效率。Message 类的属性如表 9-1 所示。

表 9-1 Message 类的属性

属　　性	描　　述
int arg1	存放一个 int 类型的数据
int arg2	存放一个 int 类型的数据
int what	存放一个 int 类型的数据，该数据表示信息的类型，用来区别其他消息
Object obj	存放任意类型的对象

9.2.3 介绍消息处理类

Handler 即为消息处理类，在 Android 中，Handler 机制主要用作异步消息处理，这是谷歌设计的一套机制，能帮助开发者有序地处理异步操作。它具有两个主要作用：实现延迟执行messages或runnables,将A线程的操作入队到B线程中。Handler类的常用方法如表9-2所示。

表9-2 Handler类的常用方法

方法名称	使用说明
post(Runnable)	即刻发送Runnable对象,并将其最后封装为Message对象
postAtTime(Runnable,long)	在指定时间内发送Runnable对象,并将其最后封装为Message对象
postDelayed(Runnable long)	延迟一定时间发送Runnable对象,并将其最后封装为Message对象
sendEmptyMessage(int)	发送空消息
sendMessage(Message)	即刻发送消息
sendMessageAtTime(Message,long)	在指定时间内发送消息
sendMessageDelayed(Message,long)	延迟一定时间发送消息

9.3 项目运行

9.3.1 创建一个线程

1. Thread类启动线程

【例9.1】 在Android Studio中创建名为9.1的App项目,然后创建一个新线程用来播放音频文件,并实现在音频播放完毕后每3秒钟循环播放。

步骤1:在Android Studio中创建一个名为9.1的App项目,打开res下layout文件夹中的activity_main.xml文件,在布局管理器中添加START按钮,用于控制音频的播放。其具体代码如下:

视频讲解

```
<?xml version = "1.0" encoding = "utf-8"?>
< LinearLayout xmlns:android = "http://schemas.android.com/apk/res/android"
    android:layout_width = "fill_parent"
    android:layout_height = "fill_parent" >

    < Button
```

```
        android:id = "@ + id/button"
        android:layout_width = "wrap_content"
        android:layout_height = "wrap_content"
        android:text = "@string/startbutton" />
</LinearLayout>
```

步骤 2：在 MainActivity.java 文件中对播放音乐的线程进行定义，并指定音频位置。其关键代码如下：

```
public class MainActivity extends AppCompatActivity {
    private Thread thread;                              // 定义一个线程
    private static MediaPlayer music = null;            // 定义一个播放器
    private static String musicPath;                    // 定义音频路径

    @Override
    public void onCreate(Bundle savedInstanceState) {
        super.onCreate(savedInstanceState);
        setContentView(R.layout.activity_main);

        Uri uri = MediaStore.Audio.Media.INTERNAL_CONTENT_URI;   // 指定播放音频的位置
        String[] projection = {
                MediaStore.Audio.Media.DATA
        };
        Cursor cursor = getContentResolver().query(uri,projection,null,null,null);
        if(cursor.moveToFirst()){
            musicPath = cursor.getString(0);
        }
```

步骤 3：在 MainActivity.java 文件中编写播放音频的方法，以便按钮事件调用。其关键代码如下：

```
private void startmusic() {
        if (music != null) {
            music.release();
        }
        music = new MediaPlayer();
        try {
            music.setDataSource(musicPath);
            music.prepare();
        } catch (IOException e) {
            e.printStackTrace();
        }
        music.start();
            music.setOnCompletionListener(new OnCompletionListener() {

            @Override
```

```
            public void onCompletion(MediaPlayer music) {
                try {
                    Thread.sleep(3000);
                    startmusic();
                } catch (InterruptedException e) {
                    e.printStackTrace();
                }

            }
        });
    }
```

步骤4：继续在 MainActivity.java 文件中设置按钮事件，当单击按钮时调用上述播放音频的方法 startmusic()。其关键代码如下：

```
        Button newbutton = (Button) findViewById(R.id.button);
        newbutton.setOnClickListener(new OnClickListener() {
            @Override
            public void onClick(View v1) {
                ((Button) v1).setEnabled(false);
                thread = new Thread(new Runnable() {
                    @Override
                    public void run() {
                        startmusic();

                    }
                });
                thread.start();
            }
        });
    }
```

步骤5：在 MainActivity.java 文件中设置音频播放完毕结束进程的方法。其关键代码如下：

```
    @Override
    protected void over() {
        if (music != null) {
            music.stop();
            music.release();
            music = null;
        }
        if (thread != null) {
            thread = null;
```

```
        }
        super.over();
    }
}
```

步骤6：运行程序，效果如图9-1所示。

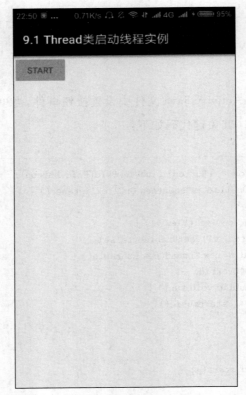

图9-1 Thread类启动一个线程的运行界面

2. Thread 类中断线程

【例9.2】 在 Android Studio 中创建名为 9.2 的 App 项目，实现一个新建线程的启动与中断功能。

步骤1：在 Android Studio 中创建一个名为 9.2 的 App 项目，打开 res 下 layout 文件夹中的 activity_main.xml 文件。为了能够实现对线程的启动和中断控制，可以在布局管理器中添加 START 和 STOP 两个按钮。其具体代码参照例 9.1。

步骤2：在 MainActivity.java 文件中创建一个 Runnable 对象，并重写 run()方法，用于判断线程的运行情况。其关键代码如下：

```
public class MainActivity extends AppCompatActivity implements Runnable {
    private Thread thread;
    int i;
    @Override
    public void run() {
        while(!Thread.currentThread().isInterrupted()){
            i++;
            Log.i("启动线程标识: ",String.valueOf(i));
        }
    }
}
```

步骤 3：在 MainActivity.java 文件中设置 START 按钮和 STOP 按钮的事件。其中，button1 对应 START 按钮；button2 对应 STOP 按钮。START 按钮事件的具体代码参见例 9.1，STOP 按钮事件的关键代码如下：

```
Button stopmusic = (Button)findViewById(R.id.button2);
stopmusic.setOnClickListener(new OnClickListener() {
    @Override
    public void onClick(View v) {
        if(thread!= null){
            thread.interrupt();
            thread = null;
        }
        Log.i("Tips: ","Stop the thread");   ——→ 中断线程时在日志面板中输出提示信息
    }
});
```

步骤 4：在 MainActivity.java 文件中设置音频播放完毕结束进程的方法，具体代码参见例 9.1。

步骤 5：运行程序，效果如图 9-2 所示，中断线程时在日志面板中输出的提示信息如图 9-3 所示。

9.3.2 添加消息类

【例 9.3】 在 Android Studio 中创建名为 9.3 的 App 项目，然后创建一个继承 Thread 类的新线程，并添加一个 Handler 对象处理消息。

步骤 1：在 Android Studio 中创建一个名为 9.3 的 App 项目。为了创

视频讲解

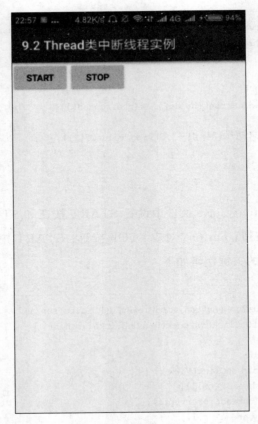

图 9-2　Thread 类中断一个线程的运行界面

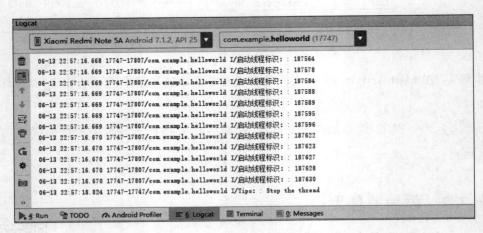

图 9-3　在日志面板中输出的"中断线程标识"的运行界面

建继承 Thread 类的新线程,在 MainActivity.java 的同级目录下新建名为 newThread 的 Java 文件。其具体代码如下:

```
package com.example.helloworld;
import android.os.Handler;
import android.os.Looper;
import android.os.Message;
import android.util.Log;
public class newThread extends Thread {          → 创建newThread类
    public Handler newhandler;                    → 创建Handler对象
    @Override
    public void run() {
        super.run();
        Looper.prepare();                         → 初始化Looper对象
        newhandler = new Handler() {
            public void handleMessage(Message msg) {
                Log.i("Looper", String.valueOf(msg.what));   → 传递消息的进程在日志中输出的提示信息
            }
        };
        Message newmessage = newhandler.obtainMessage();
        newmessage.what = 1234;                   → 传递消息的内容
        newhandler.sendMessage(newmessage);
        Looper.loop();
    }
}
```

步骤2：在 MainActivity.java 文件中引入前面创建的 newThread 线程，并将该线程启动。其具体代码如下：

```
package com.example.helloworld;
import android.app.Activity;
import android.os.Bundle;
import android.support.v7.app.AppCompatActivity;
public class MainActivity extends AppCompatActivity {
    @Override
    public void onCreate(Bundle savedInstanceState) {
        super.onCreate(savedInstanceState);
        setContentView(R.layout.activity_main);
        newThread thread = new newThread();
        thread.start();                           → 启动线程
    }
}
```

步骤3：运行程序，效果如图 9-4 所示，传递消息时在日志面板上输出的提示消息如图 9-5 所示。

图 9-4 Handler 类添加消息类的运行界面

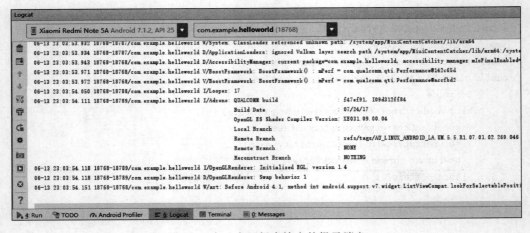

图 9-5 在日志面板中输出的提示消息

9.3.3 添加消息处理类

【例 9.4】 在 Android Studio 中创建名为 9.4 的 App 项目，制作打地鼠游戏。

视频讲解

步骤 1：在 Android Studio 中创建一个名为 9.4 的 App 项目。在 res 下 layout 文件夹中的 activity_main.xml 文件中添加打地鼠游戏所用的背景图片和地鼠图片，所用图片已经放置在资源文件夹中。其具体代码如下：

```xml
<?xml version = "1.0" encoding = "utf-8"?>
<LinearLayout xmlns:android = "http://schemas.android.com/apk/res/android"
    android:id = "@+id/fl"
    android:orientation = "vertical"
    android:background = "@drawable/background"     ——→ 添加背景图片
    android:layout_width = "match_parent"
    android:layout_height = "match_parent">
    <ImageView
        android:visibility = "gone"
        android:id = "@+id/imageView1"
        android:layout_width = "50dp"
        android:layout_height = "50dp"
        android:src = "@drawable/mouse" />           ——→ 添加地鼠图片
</LinearLayout>
```

步骤 2：根据背景图片的地洞位置，用二元数组的方式在 MainActivity.java 文件中标识出这些位置的坐标，并记录地鼠出现的位置。其关键代码如下：

```java
private void initPosition() {

    WindowManager wm = (WindowManager)
                this.getSystemService(Context.WINDOW_SERVICE);
    int width = wm.getDefaultDisplay().getWidth();
    int height = wm.getDefaultDisplay().getHeight();

    int x1 = (int)(width * 1.0/6);
    int x2 = (int)(width * 4.0/9) - 20;
    int x3 = (int)(width * 5.0/7) - 20;
    int y1 = (int)(height * 9.0/26) - 70;
    int y2 = (int)(height * 1.0/2) - 80;
    int y3 = (int)(height * 5/7) - 100;

    position = new int[][]{
            {x1,y1},{x2,y1},{x3,y1},
```

```
                    {x1,y2},{x2,y2},{x3,y2},
                    {x1,y3},{x2,y3},{x3,y3}
        };
        String s  = position[0][0] + "," + position[0][1];
        String s1 = position[1][0] + "," + position[1][1];
        String s2 = position[2][0] + "," + position[2][1];
        String s3 = position[3][0] + "," + position[3][1];
        String s4 = position[4][0] + "," + position[4][1];
        String s5 = position[5][0] + "," + position[5][1];
        String s6 = position[6][0] + "," + position[6][1];
        String s7 = position[7][0] + "," + position[7][1];
        String s8 = position[8][0] + "," + position[8][1];

        Log.d(" ----- ", "initPosition: 屏幕宽高: width = " + width + ", height = " + height );
        Log.d(" ----- ", "initPosition: (" + s + ")(" + s1 + ")(" + s2 + ")");
        Log.d(" ----- ", "initPosition: (" + s3 + ")(" + s4 + ")(" + s5 + ")");
        Log.d(" ----- ", "initPosition: (" + s6 + ")(" + s7 + ")(" + s8 + ")");
    }
```

步骤3：地鼠移动的坐标即为一个消息，消息类需要在 MainActivity.java 文件中传递。其具体代码如下：

```
        handler = new Handler() {
            @Override
            public void handleMessage(Message newmsg) {
                int index = 0;
                if (newmsg.what == 1234) {
                    index = newmsg.arg1;
                    mouse.setX(position[index][0]);
                    mouse.setY(position[index][1]);
                    mouse.setVisibility(View.VISIBLE);
                }
                super.handleMessage(newmsg);
            }
        };
        Thread hitmouse = new Thread(new Runnable() {
            @Override
            public void run() {
                int index = 0;
                while (!Thread.currentThread().isInterrupted()) {
                    index = new Random().nextInt(position.length);
                    Message m1 = handler.obtainMessage();
                    m1.what = 1234;
```

→ 记录地鼠位置的索引值index

→ 创建打地鼠线程

→ 产生地鼠的随机位置

```
                    m1.arg1 = index;
                    handler.sendMessage(m1);          ──→ 传递地鼠位置信息
                    try {
                    Thread.sleep(new Random().nextInt(400) + 400);
                } catch (InterruptedException e) {
                        e.printStackTrace();
                    }
                }
            }
        });
        hitmouse.start();                              ──→ 启动线程

}
```

步骤 4：在 MainActivity.java 文件中编写代码，实现在界面上显示打到地鼠的数量。其关键代码如下：

图 9-6　打地鼠游戏运行界面

```
mouse = (ImageView) findViewById(R.id.imageView1);
        mouse.setOnTouchListener(new OnTouchListener() {
            @Override
            public boolean onTouch(View v, MotionEvent event) {
                v.setVisibility(View.INVISIBLE);
                i++;
                Toast.makeText(MainActivity.this, "Hit [ " + i + " ]mouse!",
                        Toast.LENGTH_SHORT).show();
                return false;
            }
        });
```

步骤 5：运行程序，效果如图 9-6 所示。

9.4　项目结案

本项目为大家介绍了如何实现 Android 的多线程操作，包括如何创建一个线程，如何启动、休眠和中断线程，以及在线程运行的过程中如何实现消息的传递和处理。这些应用在 App 的内部运行机制中相当重要，尤其在游戏 App 开发过程中更为关键。大家需要不断地认知与练习才能够体会到线程的内涵与实现方法。

9.5 项目练习

1. 设计一款 App，实现一个线程的简单运行，并将运行消息输出在日志面板上。
2. 设计一款 App，实现界面的延时替换。
3. 设计一款 App，实现消息的传递，并将消息呈现在手机界面上。
4. 设计一款 App，实现一个红色方形的水平运动。

项目10

发布App

10.1 项目目标：打包与发布 App

通过前面项目的学习，大家已经可以做出具有交互效果的 App 项目了，但是仅限于通过模拟器运行。如果希望更多的用户能够通过手机来应用和体验，则需要把 App 项目打包和发布，用户下载相关文件后安装便可以运行。本项目将通过实际的案例为大家清晰地讲解如何通过 Android Studio 为 App 打包、签名、发布。

10.2 项目准备

10.2.1 介绍 META-INF 文件夹

在开发完一款 App 之后，需要对其进行打包，才能够发布供用户使用的程序。在 Android Studio 中已经内置了打包的工具。一般而言，可以通过两种方式进行打包：一是

通过 Gradle 配置打包；二是通过 Build|Generate Signed APK 打包。在打包之后生成已经签名的 APK 文件，修改其扩展名为.zip，然后进行解压操作。在解压文件中能够看到 META-INF 文件夹，里面有签名验证的文件，其中包括 MANIFEST.MF、CERT.SF 和 CERT.RSA 文件，它们分别保存着不同的签名信息。

打开 MANIFEST.MF 文件，其具体代码如下：

```
Manifest-Version: 1.0
Created-By: 1.0 (Android)

Name: res/layout/activity_main
SHA1-Digest: TKJzyMwELyakLZYM83o10LERyPQ=

Name: AndroidManifest.xml
SHA1-Digest: vf51A+/qPTUhmRyQmU6GS83eO9Y=

Name: res/drawable-v24/keys.png
SHA1-Digest: 3nPhCCVKGHdAha70YYcNvESbv5g=

Name: resources.arsc
SHA1-Digest: uh4vliR9xNyjDpU3d+WmfzTIumE=

Name: classes.dex
SHA1-Digest: S83QHv3CvsRo3e4gWNpZpvifWzg=
```

从上述代码中可以看到 App 中的每个文件下面都对应着一个 SHA1-Digest 值，这个值即为该文件进行 base64 编码后的结果。

打开 CERT.SF 文件，其具体代码如下：

```
Signature-Version: 1.0
Created-By: 1.0 (Android)
SHA1-Digest-Manifest: Uin+pH/oQLOt1Esnw9TTJpf8URc=

Name: res/layout/activity_main
SHA1-Digest: +zm+W/d5nXnQRHhQq1BeXsj4sWU=

Name: res/drawable-v24/keys.png
SHA1-Digest: 9CMNr6u3Zg/XymrpDC4NH/Qb+GE=

Name: AndroidManifest.xml
SHA1-Digest: q4qz8AP4LsfMh0TWEgTcSif6eqg=

Name: resources.arsc
SHA1-Digest: U1T+Km9u0pHDYncmJTz+Fae35iU=
```

```
Name: classes.dex
SHA1 - Digest: iOqu/znF0ISqd6UtTmA4d5isoQs =
```

从上述代码中可以看到多了一项 SHA1-Digest-Manifest 的值,这个值就是 MANIFEST.MF 文件的 SHA-1 在 base64 编码后的值。

最后一个文件 CERT.RSA 中则包含了公钥信息和发布机构的信息。

10.2.2 介绍 jar 包与 arr 包

在 Android 中,为了完成 App 的更新工作,可以对资源和代码进行打包。一般而言,在 Android Studio 中包含两种打包方式,分别是 jar 包和 arr 包。二者之间的区别十分明显。jar 包只包含了 class 文件与清单文件,不包含资源文件,例如图片等 res 文件夹中包含的所有资源。所以,如果要使用 jar 包里的资源就要使用反射来实现,否则就会出现错误。在打包 arr 的时候,Android Studio 会自动将资源文件和源代码一起打进去,这样用户在使用的时候就不用担心资源缺失的问题了。

1. jar 包的生成

首先,将待打包的 App 项目文件夹下的 build.gradle 文件进行修改,即确保该 App 处于 library 状态。其具体代码如下:

```
apply plugin: 'com.android.library
```

然后,在 build.gradle 文件中添加下列代码:

```
task clearJar(type: Delete) {
    delete 'build/makenewjar.jar'
}
task makeJar(type:org.gradle.api.tasks.bundling.Jar) {
    baseName 'makenewjar'                              →确定生成的jar包的名字
    from('build/android-profile/')                     →打包文件的位置
    into('build/')                                      →打包到的位置
    exclude('example/', 'BuildConfig.class', 'R.class') →除去不需要打包的路径和文件
    }
}
```

最后,单击 Android Studio 界面右侧的 Gradle 面板,在项目或者该类库的目录中找到 Tasks|other|makeJar,双击 makeJar 选项之后,稍等片刻即可生成 jar 包。

2. arr 包的生成

将 App 项目打包成 arr 包是 Android Studio 的新特性,这种打包方式可以将 App 项目中所使用的资源文件一起打包,完整且准确。

首先,在 App 项目中新建一个 library。

然后,在项目中依赖这个 library,且直接运行。即在 build.gradle 文件中定义属性,其具体代码如下:

```
apply plugin: 'com.android.library
```

最后,在 build/outputs 目录下生成两个 arr 包。

10.2.3 介绍 App 如何上线

当一款 App 设计完成时,可以将其发布到开放的应用商店上,以供需要该功能的用户下载、安装和使用。目前针对安卓系统的应用商店有很多,包括各种品牌手机自带的应用商店等。将 App 上线的方法十分简单,这里仅以应用宝为例,展示 App 上线的具体流程。

首先,在浏览器中输入"腾讯开放平台",打开其首页后找到"应用开放平台"入口,并单击进入,如图 10-1 所示。

图 10-1 "应用开放平台"入口

然后,选择左侧的"应用接入指引",在其右侧将展开关于 App 上线的不同流程,用户可根据自己的实际情况进行选择,如图 10-2 所示。在一般情况下,用户应首先选择"开发者注册"这一项,进而将自己的信息注册于该平台。

接着,新用户创建应用、申请资源、完善应用信息、提交上线申请,继而完成 App 的上线流程,等待平台审核通过。

图 10-2 应用接入指引界面

10.3 项目运行

视频讲解

【例 10.1】 使用例 9.4 的项目实现 App 打包与签名。

步骤 1:创建签名文件。在例 9.4 的基础上对该项目进行签名,类似身份证一样,标识出唯一版本及相关信息。在菜单栏中选择 Build|Generate Signed APK 选项,如图 10-3 所示。

步骤 2:在 Generate Signed APK 对话框中填写打包信息。由于是第一次打包,所以单击 Create new 按钮,如图 10-4 所示。

步骤 3:进入 New Key Store 对话框,填写空白项,如图 10-5 所示。其中,Key store path 列表框中一般选择本项目所在的路径,以确保其唯一性;Alias 文本框中填写该密钥的别名;Validity 列表框中填写密钥的使用年限;Certificate 选项组为开发者信息,选填即可。

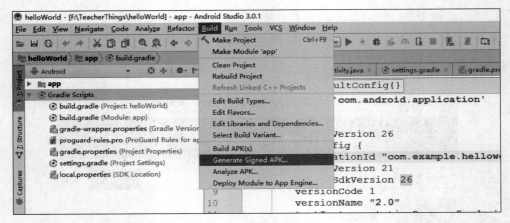

图 10-3 创建签名文件

图 10-4 Generate Signed APK 对话框

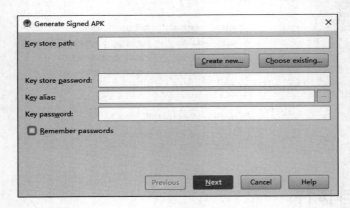

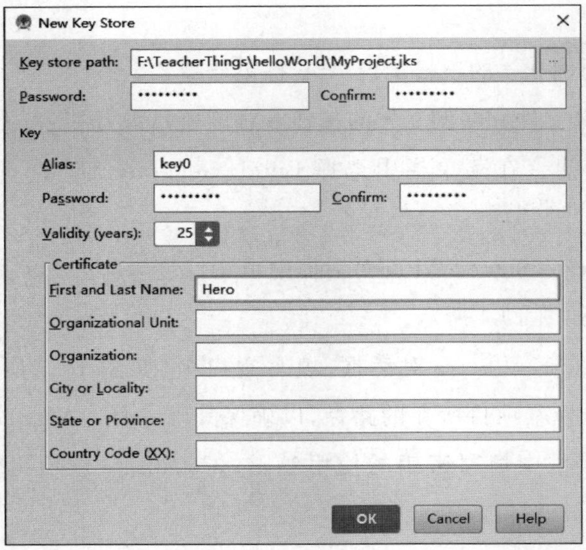

图 10-5 New Key Store 对话框

步骤4：单击 OK 按钮，返回至 Generate Signed APK 对话框。为了便于密码的记忆，可选中 Remember passwords 复选框，如图 10-6 所示。

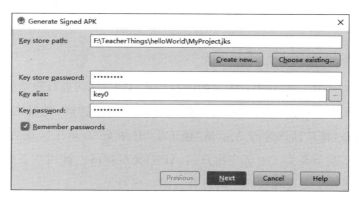

图 10-6　选中 Remember passwords 复选框

步骤5：单击 Next 按钮，在 Generate Signed APK 对话框中出现 Signature Versions（签名版本号）。这是从 Android 7.0 版本以后引入的一个新的签名机制，它为 APK 附加了一些特性使其更具安全性，所以这个选项虽然不是强制性的，但最好两个都选，如图 10-7 所示。如果选择 V2 会产生错误，则可以不选 V2。

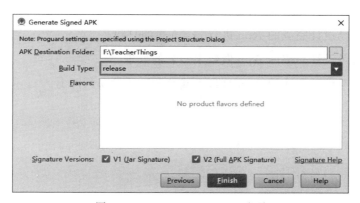

图 10-7　Signature Versions 选项

步骤6：单击 Finish 按钮，就可以在本项目路径下找到已经打包成功的 APK 文件，如图 10-8 所示。

图 10-8　APK 文件

10.4 项目结案

本项目详细地介绍了App项目如何能够打包成APK,如何进行数字签名,如何生成jar包、arr包等实际操作。通过以上内容的分析与实际案例的讲解,相信大家终于能够实现App的打包与发布,将自己设计的App从"独乐乐"升华至"众乐乐",让自己的设计被更多的人所接触、关注,甚至喜爱。相信当用户将APK发布到网上被用户下载并安装时,心中的激动和喜悦会将在学习过程中产生的所有苦涩一扫而光。

10.5 项目练习

1. 将一款已经设计完成的App进行jar方式的打包。
2. 将一款已经设计完成的App进行arr方式的打包。
3. 将一款已经设计完成的App做数字签名的处理。
4. 为一款已经设计完成的App打包生成APK,并发布到网上。

综合案例

综合案例一：猜数字

视频讲解

案例简介：设计一款关于猜数字的小游戏 App，当游戏运行时自动生成 1～100 的任一数字，玩家可以在指定位置输入所猜数字，游戏会帮助玩家判别所猜数字与生成数字之间的大小比较。每当玩家猜对时奖励 10 分，猜错一次扣除 1 分。

运行结果：如图 A-1～图 A-3 所示。

综合案例二：闹钟

视频讲解

案例简介：设计一款闹钟 App，要求能够显示当前的时间，能够设定 3 个闹钟，同时还能够删除已经设定的闹钟。

运行结果：如图 A-4～图 A-6 所示。

综合案例三：歌库

视频讲解

案例简介：设计一款歌库 App，用于整理自己的歌单，用户可以将歌手的名字、歌曲的名称输入到指定的位置进行保存，同时可以将已有歌库条目删除。

运行结果：如图 A-7～图 A-11 所示。

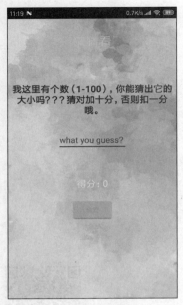

图 A-1 游戏开始界面

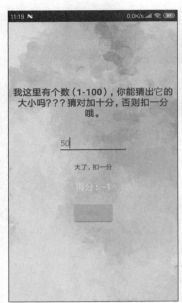

图 A-2 玩家猜错界面

图 A-3 玩家猜对界面

图 A-4 闹钟 App 初始界面

图 A-5 设置闹钟的界面

图 A-6 闹钟响铃的界面

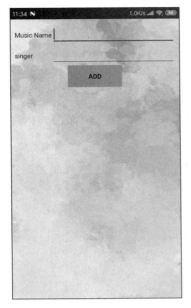

图 A-7　App 初始界面　　　图 A-8　添加信息界面　　　图 A-9　歌库列表信息界面

图 A-10　删除歌库条目　　　图 A-11　删除成功界面

图书资源支持

感谢您一直以来对清华版图书的支持和爱护。为了配合本书的使用,本书提供配套的资源,有需求的读者请扫描下方的"书圈"微信公众号二维码,在图书专区下载,也可以拨打电话或发送电子邮件咨询。

如果您在使用本书的过程中遇到了什么问题,或者有相关图书出版计划,也请您发邮件告诉我们,以便我们更好地为您服务。

我们的联系方式:

地　　址: 北京市海淀区双清路学研大厦 A 座 701

邮　　编: 100084

电　　话: 010-83470236　010-83470237

资源下载: http://www.tup.com.cn

客服邮箱: 2301891038@qq.com

QQ: 2301891038(请写明您的单位和姓名)

用微信扫一扫右边的二维码,即可关注清华大学出版社公众号"书圈"。

资源下载、样书申请
书圈

扫一扫,获取最新目录

课程直播